ORBITAL DEBRIS

A Technical Assessment

Committee on Space Debris

Aeronautics and Space Engineering Board

Commission on Engineering and Technical Systems

National Research Council

NATIONAL ACADEMY PRESS
Washington, D.C.   1995

titution Ave., N.W. ▪ Washington, DC 20418

Orbital debris

of this report was approved by the Governing
whose members are drawn from the councils of
the National Academy of Sciences, the National Academy of Engineering, and the Institute
of Medicine. The members of the committee responsible for the report were chosen for
their special competencies and with regard for appropriate balance.

This report has been reviewed by a group other than the authors according to procedures
approved by a Report Review Committee consisting of members of the National Academy
of Sciences, the National Academy of Engineering, and the Institute of Medicine.

This study was supported by Grant NAGW 3414 between the National Academy of
Sciences and the National Aeronautics and Space Administration.

Library of Congress Cataloging-in-Publication Data

National Research Council (U.S.). Committee on Space Debris.
    Orbital debris : a technical assessment / Committee on Space
Debris, Aeronautics and Space Engineering Board, Commission on
Engineering and Technical Systems, National Research Council.
        p.   cm.
    Includes bibliographical references and index.
    ISBN 0-309-05125-8
    1. Space debris. I. Title.
TL1499.N38   1995
629.4'16—dc20   95-18686
                                                                      CIP

Cover Illustration: Location of cataloged space objects as of 9:12 p.m. on December 15, 1993.
Each dot represents one cataloged space object. Source: Prepared by Kaman Sciences
Corporation based on U.S. Space Command Satellite Catalog.

iii

Bradford W. Parkinson, Professor, Aeronautics and Astronautics and Hansen Experimental Physics Laboratory, Stanford University, High Energy Physics Laboratory, Stanford, California

Alfred Schock, Director, Energy System Department, Orbital Sciences Corporation, Germantown, Maryland

John D. Warner, President, Boeing Computer Services, Seattle, Washington

ASEB Staff

JoAnn C. Clayton, *Director*
Alan C. Angleman, *Senior Program Officer*
Allison C. Sandlin, *Senior Program Officer*
Noel E. Eldridge, *Program Officer*
Paul J. Shawcross, *Program Officer*
Anna L. Farrar, *Administrative Associate*
Wlliam E. Campbell, *Administrative Assistant*
Mary T. McCormack, *Senior Project Assistant*
Ted W. Morrison, *Program Assistant*
Beth A. Henry, *Project Assistant*

The National Academy of Sciences is a private, nonprofit, self-perpetuating society of distinguished scholars engaged in scientific and engineering research, dedicated to the furtherance of science and technology and to their use for the general welfare. Upon the authority of the charter granted to it by the Congress in 1863, the Academy has a mandate that requires it to advise the federal government on scientific and technical matters. Dr. Bruce Alberts is president of the National Academy of Sciences.

The National Academy of Engineering was established in 1964, under the charter of the National Academy of Sciences, as a parallel organization of outstanding engineers. It is autonomous in its administration and in the selection of its members, sharing with the National Academy of Sciences the responsibility for advising the federal government. The National Academy of Engineering also sponsors engineering programs aimed at meeting national needs, encourages education and research, and recognizes the superior achievements of engineers. Dr. Robert M. White is president of the National Academy of Engineering.

The Institute of Medicine was established in 1970 by the National Academy of Sciences to secure the services of eminent members of appropriate professions in the examination of policy matters pertaining to the health of the public. The Institute acts under the responsibility given to the National Academy of Sciences by its congressional charter to be an adviser to the federal government and, upon its own initiative, to identify issues of medical care, research, and education. Dr. Kenneth I. Shine is president of the Institute of Medicine.

The National Research Council was organized by the National Academy of Sciences in 1916 to associate the broad community of science and technology with the Academy's purposes of furthering knowledge and advising the federal government. Functioning in accordance with general policies determined by the Academy, the Council has become the principal operating agency of both the National Academy of Sciences and the National Academy of Engineering in providing services to the government, the public, and the scientific and engineering communities. The Council is administered jointly by both Academies and the Institute of Medicine. Dr. Bruce Alberts and Dr. Robert M. White are chairman and vice-chairman, respectively, of the National Research Council.

Preface

Over the last 37 years, thousands of spacecraft have been launched into orbit for scientific, commercial, environmental, and national security purposes. One consequence of this activity has been the creation of a large population of debris—artificial space objects that serve no useful function—in orbit around the Earth. Much of this debris will remain in orbit for hundreds of years or more, posing a long-term hazard to future space activities. Currently, the hazard is fairly low; there are no confirmed instances of orbital debris seriously damaging or destroying a spacecraft. However, continuing space operations and collisions between objects already in orbit are likely to generate additional debris faster than natural forces remove it, potentially increasing the debris hazard in some orbital regions to levels that could seriously jeopardize operations in those regions.

To acquire an unbiased technical assessment of (1) the research needed to better understand the debris environment, (2) the necessity and means of protecting spacecraft against the debris environment, and (3) potential methods of reducing the future debris hazard, the National Aeronautics and Space Administration asked the National Research Council to form an international committee to examine the orbital debris issue. The committee was asked to draw upon available data and analyses to

- characterize the current debris environment,
- project how this environment might change in the absence of new measures to alleviate debris proliferation,
- examine ongoing alleviation activities,

- explore measures to address the problem, and
- develop recommendations on technical methods to address the problems of debris proliferation.

In the summer of 1993, the National Research Council formed a committee of 11 technical experts from six spacefaring nations to perform this task. This report, which draws upon existing research, the expertise of committee members, and material presented in the study's November 1993 workshop, represents that committee's consensus view.

The committee strove to ensure that the study focused on technical issues. This report does not suggest appropriate funding levels for future debris research, propose specific protective measures for particular spacecraft, or lay out detailed implementation strategies for techniques to contain the future debris hazard. Decisions on such matters involve political and economic as well as technical considerations and must be made by entities capable of weighing all these factors. Rather, this report seeks to provide engineers, scientists, and policy makers with the sound technical information and advice upon which such decisions must be based.

The committee would like to thank the many experts who briefed the committee, participated in the study's workshop, or in other ways helped us over the course of this study. I would like to personally thank the members of the committee for their hard work and dedication in developing this report. Finally, this project could not have been completed without the dedication and efficiency of the staff of the Aeronautics and Space Engineering Board. In particular, I want especially to thank Paul Shawcross, the Study Director, whose hard work, technical knowledge, organizational skills, writing and editing ability, and ever-present positive attitude have been key to a successful outcome.

George Gleghorn
Chair

Contents

ix

ORBITAL DEBRIS

Executive Summary

Space activities in Earth orbit are increasingly indispensable to our civilization. Orbiting spacecraft serve vital roles as communications links, navigation beacons, scientific investigation platforms, and providers of remote sensing data for weather, climate, land use, and national security purposes. The spacecraft that perform these tasks are concentrated in a few orbital regions, including low Earth orbit (LEO), semisynchronous orbit, and geosynchronous Earth orbit (GEO). These orbital regions represent valuable resources because they have characteristics that enable spacecraft operating within them to execute their missions more effectively.

Functional spacecraft share the near-Earth environment with natural meteoroids and the orbital debris that has been generated by past space activities. Meteoroids orbit the Sun and rapidly pass through and leave the near-Earth region (or burn up in the Earth's atmosphere), resulting in a fairly continual flux of meteoroids on spacecraft in Earth orbit. In contrast, artificial debris objects (including nonfunctional spacecraft, spent rocket bodies, mission-related objects, the products of spacecraft surface deterioration, and fragments from spacecraft and rocket body breakups) orbit the Earth and will remain in orbit until atmospheric drag and other perturbing forces eventually cause their orbits to decay into the atmosphere. Since atmospheric drag decreases as altitude increases, large debris in orbits above about 600 km can remain in orbit for tens, thousands, or even millions of years.

Although the uncontrolled reentry of some orbital debris could potentially pose a hazard to activities on the Earth's surface, the major

1

hazard posed by debris is to space operations. Although the current hazard to most space activities from debris is low, growth in the amount of debris threatens to make some valuable orbital regions increasingly inhospitable to space operations over the next few decades. A responsible approach to orbital debris will require continuing efforts to increase our knowledge of the current and future debris population, the development of tools to aid spacecraft designers in protecting spacecraft against the debris hazard, and international implementation of appropriate measures to minimize the creation of additional debris.

CHARACTERIZING THE DEBRIS ENVIRONMENT

The debris environment is difficult to characterize accurately. First, the debris population changes continually as new debris is created and existing debris reenters the Earth's atmosphere. Detection of such changes requires that measurements of the debris environment be updated periodically. Second, only the largest objects can be repeatedly tracked by ground-based sensors; tracking of the numerous smaller pieces of debris is much more difficult. The U.S. and Russian space surveillance systems are able to track and catalog virtually all objects larger than 20 cm diameter in LEO. However, as altitude increases, the minimum-sized object that these systems are capable of tracking increases, until at GEO only objects larger than about 1 meter in diameter are presently cataloged.

Characterization of the debris population that cannot be cataloged must thus be accomplished by sampling the orbital debris flux at particular locations and times and using these data as a basis for estimating the characteristics of the general population. The flux can be sampled either directly (with spacecraft surfaces that are struck by debris) or remotely (by using ground- or space-based radars or optical telescopes that record debris passing through their fields of view). Presently, ground-based remote sensing is the most effective method for sampling the medium-sized (approximate diameter 1 mm–10 cm) debris population, and in situ impact sampling is the most effective method for measuring the small (approximate diameter <1 mm) debris population.

Current measurements of the debris environment contain gaps, such as a lack of information on objects smaller than 1 meter in diameter in GEO, on the small debris population at altitudes above 600 km, and on the medium-sized debris population above LEO. There are, however, several promising means for better characterizing the debris population. For example, large-aperture optical telescopes or telescopes equipped with charge-coupled devices could be employed to improve cataloging of large (approximate diameter >10 cm) debris in orbits above LEO, shorter-wavelength radars situated at low latitudes could be used to improve our knowledge of the medium-sized debris population in LEO, and active impact detectors deployed at altitudes above 600 km could extend our knowledge of the distribution of the small debris population.

Using such means to better characterize the orbital debris environment and applying the knowledge thus acquired can increase the cost-effectiveness of efforts both to reduce the future debris population and to protect spacecraft from debris. This is not to suggest an effort to characterize all debris in all orbits; rather, characterization efforts should aim at providing information needed to fill critical gaps in the data. To focus this effort *the committee recommends that an international group be formed (1) to advise the space community about areas in the orbital debris field needing further investigation and (2) to suggest potential investigation methods.*

As an interim set of debris characterization research priorities, *the committee recommends the following:*

- *models of the future debris environment should be further improved,*
- *uncataloged debris in LEO should be carefully studied,*
- *further studies should be conducted to better understand the GEO debris environment,*
- *a strategy should be developed to gain an understanding of the sources and evolution of the small debris population, and*
- *the data acquired from this research should be compiled into a standard population characterization reference model.*

To improve the efficiency of orbital debris research, *the committee recommends exploring the creation of an international system for collecting, storing, and distributing data on orbital debris.* Finally, to ensure the accuracy of the data produced by these efforts, *the committee recommends that the orbital debris community exercise more peer review over its research.*

HAZARD TO SPACE OPERATIONS FROM DEBRIS

The natural meteoroid environment does not pose a serious hazard to well-designed spacecraft in Earth orbit. However, there are now orders of magnitude more large orbital debris objects than large meteoroids in the near-Earth area at any given time. Although measurements of the medium-sized debris environment are sparse, the population of medium-sized orbital debris also appears to be greater than the population of medium-sized micrometeoroids in the regions of LEO where measurements have been made.

Spacecraft are much more likely to collide with smaller debris than with larger objects. In LEO, the probability of collision with debris in each size range is believed typically to increase by more than a factor of 100 for every factor of 10 decrease in size over most of the medium to small debris size ranges. (For example, LEO spacecraft are probably at least 100 times more likely to be struck by 1-mm-diameter objects than by 1-cm-diameter objects.) In the orbital altitude most densely populated with debris (between 900 and 1,000 km), models suggest that a typical spacecraft (10-square-meter cross-sectional area) has only about one chance in 1,000 of colliding with a large debris object over the spacecraft's 10-year functional lifetime. The chance of colliding with 1- to 10-cm debris over the same period, however, is estimated to be about 1 in 100, a collision with 1-mm to 1-cm diameter debris is believed to be likely, and frequent collisions with debris smaller than 1 mm will occur.

The chance of colliding with debris varies greatly with orbital altitude and, to a lesser extent, with orbital inclination. Based on the best available data, the probability of colliding with large or medium-sized debris in LEO is at least 100 times greater than the average probability in GEO and is likely to be 1,000 times greater than the probability in less-used orbital regions. Even within LEO, the collision probability varies greatly with altitude; for example, the chance of collision with medium-sized or large debris is probably higher by a factor of 50 at 900-km altitude than at 250 km. Measurements of small debris are so limited that it is unclear whether this population follows a similar altitude distribution.

The damage that a collision with debris can cause to a spacecraft depends on the kinetic energy released in the collision, the design of the spacecraft, and the geometry of the collision. Due to the typically high relative velocities of the objects involved, collisions in orbit can be highly energetic. For example, a 1-kg object involved in a (typical for LEO) 10-km/s collision will impact with the same relative kinetic energy (about 100 MJ) as a fully loaded 35,000-kg truck moving at 190 km/h. If the kinetic energy released in a collision is large enough compared to the mass of the objects involved, a catastrophic breakup will occur. In such a breakup, numerous fragments capable of causing further catastrophic breakups could be produced. A 1-kg object impacting at 10 km/s, for example, is probably capable of catastrophically breaking up a 1,000-kg spacecraft if it strikes a high-density element in the spacecraft. In such a breakup, numerous fragments larger than 1 kg would be created.

Even if a collision does not fragment a spacecraft, the impact may generate a variety of other damage modes (e.g., spallation, rupturing, leakage, and deformation) possibly degrading spacecraft performance or causing spacecraft failure. In LEO, debris as small as a few millimeters in diameter can puncture unprotected fuel lines and damage other sensitive components, and debris smaller than 1 mm in diameter can erode thermal surfaces and optics. The effect of debris impact on a particular space-craft is strongly dependent on the spacecraft's design; debris is far more likely to damage unprotected spacecraft than those that were designed with due consideration of the meteoroid and orbital debris environment. Components that are difficult to protect from debris (including photovol-taic arrays, suites of communications antennas, and sensors) may, how-ever, be at risk even in a well-designed spacecraft.

Assessments of the damage caused by debris impact are needed to (1) design spacecraft components and shielding capable of surviving de-bris impact, and (2) better understand the effect of collisions on the evo-lution of the future debris population. Since it is very difficult to gather data from the rare impacts of medium-sized or large debris in space, assessment of the potential damage such impacts can cause is accom-plished primarily through ground-based experimental testing and ana-lytic/numeric methods. Experimental hypervelocity impact testing gen-erally provides the majority of information for such assessments; analytic and numeric tools currently mainly supplement and extend experimen-tal results.

Current hypervelocity impact facilities cannot, however, simulate all relevant debris compositions, shapes, and velocities, and data on the vul-nerability of different spacecraft components to debris impact are lim-ited. Although analytical and numerical techniques can be used to pre-dict impact damage for regimes that hypervelocity testing cannot simulate, these predictions can be inaccurate unless they are based on experimental data. Unfortunately, many of the experimental data are not available due to the general inaccessibility of hypervelocity facility capa-bilities and the impact data generated at these facilities. As a result of these limitations, current spacecraft protection systems may not provide their desired level of protection, and current models of the effects of colli-sions on the future debris population may be inaccurate.

To facilitate the development of improved models of debris impact damage and enable the development of improved debris shielding, *the committee recommends the continuation of research to characterize the effects of hypervelocity impacts in the following areas:*

- *further development of techniques to launch projectiles to the velocities typical of collisions in LEO;*
- *improved models of the properties of new spacecraft materials;*
- *studies of impact damage effects on critical spacecraft components;*
- *development of analytical tools consistent over a range of debris impact velocities, shapes, and compositions; and*
- *improved models of catastrophic space object breakup due to debris impact.*

To reduce duplication of effort and speed this research, *a handbook describing the capabilities of international hypervelocity testing facilities should be developed.* This handbook would serve to increase opportunities for sharing data generated at different facilities.

DESIGNING FOR THE DEBRIS ENVIRONMENT

Although large uncertainties remain, an improved understanding of the debris environment, combined with (1) the growing availability of analytic and experimental tools to quantify the threat to a spacecraft from debris and (2) the development of techniques to protect against debris impacts, has made it feasible for designers to assess the debris hazard and attempt to protect their spacecraft appropriately.

Most spacecraft designers are, however, unaware of these tools and techniques, and very few understand all of the assumptions on which they are based. Such understanding is important because these tools and techniques may incorporate models that have not yet been clearly validated. For this reason, *the committee recommends that a guide for spacecraft designers—including information on environmental prediction, damage assessment, and passive and operational protection schemes—should be developed and distributed widely.*

A spacecraft's basic structure should be the first line of defense against the debris hazard, but if necessary, active, passive, and operational techniques can be employed to provide additional protection. Passive protection (shielding) of critical components is one viable means of protection. Shield development continues to decrease the mass of shielding required to protect against debris, though it is uncertain how well these shields will protect against the full range of debris sizes, shapes, and compositions. Operational protection schemes, such as the use of redundant components, may also be appropriate for some spacecraft. Such schemes can add weight and cost but can also protect the spacecraft against non-debris-related failures. Active protection measures, such as on-orbit collision avoidance, are typically expensive and difficult to implement effectively.

REDUCING THE FUTURE DEBRIS HAZARD

If the only additions to the future debris population were rocket bodies, nonfunctional spacecraft, mission-related debris, and the products of explosions and surface deterioration, the space object population would probably continue its roughly linear growth. However, several models that use different methodologies and different assumptions predict that collisions between space objects will add a potentially large and exponentially growing number of new debris objects to this population. Because of the numerous uncertainties involved in existing models of the debris environment, it is premature to suggest exactly when such collisional growth will begin to occur; it may already be under way, or it may not begin for several decades.

Collisional growth is most likely to occur in regions that (1) have a high debris flux, (2) do not experience a high level of atmospheric drag, and (3) have high typical collision velocities. Experts believe that two LEO regions that meet these criteria, at around 900- to 1,000-km and 1,500-km altitude, have already exceeded their "critical density," the point at which more fragments will be generated by collisions than will be removed by atmospheric drag, even if no further objects are added. Potential exponential growth in the debris population of these regions could force spacecraft designers and operators to take countermeasures against the threat posed by debris or face a heightened risk of losing spacecraft capability due to impacts. Because fragments from collisions in regions experiencing collisional growth may become widely distributed, the collision hazard may be raised even in regions not now threatened by collisional growth.

There are many possible ways to reduce the hazards posed by debris to future space operations. These include actions taken as a spacecraft enters orbit, during its operations, and after its functional lifetime. The active removal of space debris (e.g., the use of debris collection robots or "sweepers"), however, will not be an economical means of reducing the debris hazard in the foreseeable future. Design of future spacecraft and launch vehicles for end-of-life passivation, autonomous deorbiting, orbital lifetime reduction, and reorbiting are generally far more economical ways of reducing the collision hazard.

Growth in the debris hazard can be abated significantly without exorbitant costs by reducing the number of breakups of spacecraft and rocket bodies and, to a lesser degree, by ending or sharply reducing the amount of mission-related debris released during spacecraft deployment and operations. Such measures, however, will not reduce the total mass of objects in orbit. Since the total mass of objects in orbit is a key determinant of the rate of future collisional population growth, it will be necessary to take measures to remove spacecraft and rocket bodies from some crowded orbital regions at the end of their functional lifetimes in order to reduce the potential for exponential growth of the debris population.

Various methods have been proposed to remove spacecraft and rocket bodies from crowded orbital regions at the end of their functional lifetimes. In lower-altitude orbits, it is often possible to deorbit or reduce the orbital lifetime of spacecraft or rocket bodies, typically through a final propulsive maneuver. Although direct deorbiting into the Earth's atmosphere eliminates debris from orbit rapidly, it requires more fuel than maneuvering to reduce orbital lifetime. In orbital regions that are too high above the atmosphere for economical deorbiting or orbital lifetime reduction maneuvers, spacecraft or rocket bodies can be moved to "disposal orbits" a safe distance away from valuable orbital regions at the end of their functional lifetimes. Disposal orbits are not a viable alternative within LEO because perturbing forces make all such orbits unstable; objects in LEO disposal orbits will eventually cross heavily trafficked orbital regions. Neither the committee nor the wider debris community can agree on whether disposal orbits should be used by all spacecraft and rocket bodies in GEO.

As with other environmental issues, decisions on the adoption and implementation of particular debris reduction methods must balance political and economic as well as technical factors and thus must be made in forums that are capable of balancing all of these factors. Since implementation of debris mitigation measures can impose additional costs on space operations, international rules are needed to ensure that those engaging in debris mitigation activities are not penalized. For this reason, *the committee recommends that the spacefaring nations develop and implement debris reduction methods on a multilateral basis.*

Given the long development cycle for new space vehicles with debris-minimizing features, the technical development, cost–benefit assessments, and international discussions required to implement countermeasures should start as soon as possible. Although these multilateral discussions cannot be conducted on a purely technical basis, it is crucial that they be based on sound technical advice. *The committee's consensus technical assessment of the actions that should be taken to reduce future growth in the debris hazard, based on our current understanding of the debris environment and of the costs and benefits of various mitigation measures, is represented in the following recommendations:*

- *Space system developers should adopt design requirements to dissipate on-board energy sources to ensure that spacecraft or rocket bodies do not explode after their functional lifetimes.*
- *The release of mission-related objects during spacecraft deployment and operations should be avoided whenever possible.*
- *Spacecraft and rocket bodies should be designed to minimize the unintentional release of surface materials, including paint and other thermal control materials, both during and after their functional lifetimes.*
- *Intentional breakups in orbit (especially those expected to produce a large amount of debris) should be avoided if at all possible. No intentional breakups expected to produce numerous debris with orbital lifetimes longer than a few years should be conducted in Earth orbit.*

- *Spacecraft and rocket bodies in LEO and in highly elliptical orbits passing through LEO should either be removed from LEO or have their orbital lifetime reduced at the end of their functional lifetime.*
- *The use of GEO disposal orbits should be studied further. Until such studies produce a verifiably superior long-term strategy for dealing with the GEO debris hazard, operators of GEO spacecraft and rocket bodies should be encouraged to reorbit their spacecraft at the end of their functional lifetimes if they are capable of safely performing a reorbiting maneuver to a disposal orbit at least 300 km from GEO.*

The threat that orbital debris poses to international space activities is presently not large, but it may be on the verge of becoming significant. If and when it does, the consequences could be very costly—and extremely difficult to reverse. By contrast, the cost of reducing the growth of the hazard can be relatively low, involving specialized data collection and research along with cooperation and information sharing among the developers and users of space hardware. The committee believes that spacefaring nations should take judicious, timely steps now to understand the risk and agree on ways to reduce it.

Introduction

The volume of space surrounding the Earth has never been empty. Even before the 1957 launch of Sputnik, a rain of particles of various sizes passed constantly through near-Earth space. The hazard to functional spacecraft from such naturally occurring meteoroids, however, is low; simple shielding techniques can protect against the vast majority of these predominantly small particles, and the chance of a spacecraft colliding with a meteoroid large enough to cause serious damage is remote.

Since the beginning of space flight, however, the collision hazard in Earth orbit has steadily increased as the number of artificial objects orbiting the Earth has grown. More than 4,500 spacecraft have been launched into space since 1957; nearly 2,200 remain in orbit. Of these, about 450 are still functional; the rest can no longer carry out their missions and are considered debris. Nonfunctional spacecraft, however, constitute only a small fraction of the debris orbiting the Earth. They share Earth orbit with spent rocket bodies; the lens caps, bolts, and other "mission-related debris" released into space during a spacecraft's deployment and operation; aluminum oxide particles from the exhaust of solid rocket motors; paint chips from space object surfaces; and the numerous fragmentary objects generated by the more than 120 spacecraft and rocket body breakups that have occurred in orbit. Figure 1 depicts the range of objects in space, including various types of debris.

It is clear that this artificial orbital debris can potentially endanger functional spacecraft. In orbits near the Earth, colliding objects typically will have a relative velocity of more than 10 km/s. At these speeds,

11

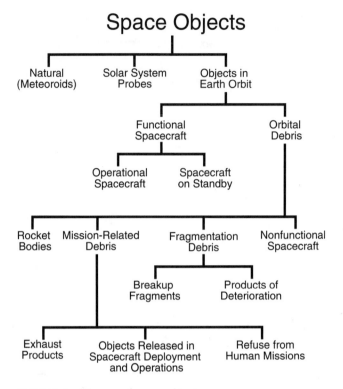

FIGURE 1   Classes of space objects.

collision with objects as small as a centimeter in diameter could damage or prove fatal to most spacecraft, depending on where the impact occurs. Impacts with the much more numerous debris particles that are a millimeter or less in diameter can damage optics, degrade surface coatings, or even crack windows.

There have not yet been any confirmed incidents in which collision with orbital debris has severely damaged or destroyed a spacecraft, but there have been a number of spacecraft malfunctions and breakups that might have been caused by impacts with debris. Smaller debris particles have certainly pitted windows (of the U.S. Space Shuttle and the Salyut and Mir space stations) and marred the surfaces of spacecraft such as the Solar Maximum Mission spacecraft (Solar Max) and the Long Duration Exposure Facility (LDEF). This type of low-grade damage is probably very widespread in low Earth orbit (LEO), but much of it goes undetected, because most spacecraft are not returned to Earth for examination.

Since the late 1970s, studies of the debris population using modeling techniques have predicted that the hazard from orbital debris is likely to grow in time unless deliberate actions are taken to minimize the creation of new debris. This predicted increased hazard will force spacecraft designers and operators to take countermeasures against the threat of debris or to face a heightened risk of losing spacecraft capability due to impacts. Projected future increases in the debris hazard already have had an effect on the design of LEO spacecraft (such as the International Space Station) that are large and have long projected functional lifetimes and, thus, a significant probability of colliding with damaging debris.

Concern about the orbital debris hazard has grown in the last decade. A number of events, including the breakup of several rocket upper stages and the replacement of Shuttle windows after impacts by small particles, helped to increase awareness of the problem, as did the need to factor space debris considerations into the design of Space Station Freedom. Reports by the American Institute of Aeronautics and Astronautics (AIAA), the European Space Agency (ESA), the U.S. Interagency Group on Space, the International Academy of Astronautics (IAA), and the Japan Society for Aeronautical and Space Sciences also served to define the problem better and to offer some suggestions on its mitigation.

Knowledge about orbital debris also has grown over the last few years. New data on the debris population have been gathered from a multitude of sources, ranging from LDEF, the European Retrievable Carrier (EURECA), and Mir—all of which collected data from the impacts of small debris in space—to the Haystack radar, which collected data on previously undetectable medium-sized debris. These new data have served to improve the models used to estimate the current characteristics and predicted growth of the overall debris population.

Despite these efforts, there remains much that we do not know about orbital debris. The primary reason is the fundamental difficulty of studying small, fast-moving, often dark objects orbiting hundreds or thousands of kilometers above the Earth. Our knowledge also is limited because

---

BOX 1 Other Effects of Orbital Debris

In addition to presenting a collision hazard to space operations, orbital debris can also have other detrimental effects. For example, debris can affect astronomical observations by leaving light trails on long-exposure photographs with wide fields of view. In addition, debris reentering the atmosphere can potentially harm people and property on the ground. In the past, this has been a minor hazard, since most reentering debris objects burn up completely in the atmosphere. However, there have been some exceptions (e.g., Kosmos 954, Skylab, and Salyut-7/Kosmos 1686), and the exact number of objects surviving reentry is unknown.

much of the data on debris to date have been collected as a by-product of non-debris-related investigations that have covered limited ranges of debris size and altitude for limited time frames. As a result, there are gaps in our understanding of the debris population: for example, estimates of the important population of LEO debris with diameters in the range of 1 mm to 10 cm still vary by a factor of two or more; we know that breakups have occurred in geosynchronous Earth orbit (GEO) only because telescopes happened to be looking at spacecraft when they broke up; most debris experts were surprised when LDEF data suggested the existence of a significant population of small debris in elliptical orbits; and there are no meaningful measurements of debris smaller than 1 mm at altitudes higher than 600 km.

Although there is still a great deal of work to be done in defining the current and future debris environment, enough data have been gathered and analyses performed that we are beginning to understand better the overall magnitude of the orbital debris problem. In addition, the broader space community is becoming aware that debris is a serious issue, and a consensus is building that actions must be taken now to preserve the space environment for the future. The challenge that we now face is to implement an appropriate set of actions to respond to the issues raised by orbital debris. This is not a simple problem, and it will not have a simple solution. A responsible approach to orbital debris will require

- continuing measurement and modeling efforts to increase our knowledge of the current and future debris population;
- the development of tools to aid spacecraft designers in protecting their spacecraft appropriately against the existing debris hazard; and
- widespread implementation of appropriate measures to minimize the creation of additional debris.

This report seeks to provide some guidance on how to achieve these goals.

BACKGROUND REFERENCES

AIAA (American Institute of Aeronautics and Astronautics). 1992. Orbital Debris Mitigation Techniques: Technical, Economic, and Legal Aspects. SP-016-1992. Washington, D.C.: AIAA.

AIAA (American Institute of Aeronautics and Astronautics). 1981. Space Debris: An AIAA Position Paper. AIAA Technical Committee on Space Systems. Washington, D.C.: AIAA.

ESA (European Space Agency). 1988. Space Debris. ESA SP-1109. Paris: ESA.

Interagency Group (Space). 1989. Report on Orbital Debris. Washington, D.C.: National Security Council.

International Academy of Astronautics Committee on Safety, Rescue, and Quality. 1992. Position Paper on Orbital Debris. August 27. Paris: International Academy of Astronautics.

JSASS (Japan Society for Aeronautical and Space Sciences). 1993. Summary of Space Debris Study Group Report. JSASS Space Debris Study Group. March 19. Tokyo: JSASS.

Simpson, J.A. 1994. Preservation of Near Earth Space for Future Generations. New York: Cambridge University Press.

U.S. Congress, House of Representatives, Committee on Science, Space, and Technology, Subcommittee on Space Sciences and Applications. 1988. Orbital Space Debris. Washington, D.C.: U.S. Government Printing Office.

U.S. Congress, Office of Technology Assessment. 1990. Orbiting Debris: A Space Environmental Problem—Background Paper OTA-BP-ISC-72. Washington, D.C.: U.S. Government Printing Office.

U.S. General Accounting Office. 1990. Space Debris a Potential Threat to Space Station and Shuttle GAO/IMTEC-90-18. Washington, D.C.: U.S. Government Printing Office.

1

Space Operations and the Space Environment

SPACE OPERATIONS

In the 37 years since the launch of Sputnik 1, space operations have become an integral component of the world's economy, scientific activities, and security systems. Orbital debris is inextricably linked with these operations: debris is created in the course of these operations and is important because it poses a potential threat to future operations. Understanding some of the characteristics of historical, current, and future space operations is thus essential to understanding the overall debris problem.

Today, spacecraft owned by 23 nations and several international organizations (representing more than 100 countries) support a wide variety of important missions, including communications, navigation, meteorology, geodesy and geophysics, remote sensing, search and rescue, materials and life sciences, astrophysics, and national security. A broad spectrum of simple and sophisticated functional spacecraft, with masses ranging from tens of kilograms to tens of metric tons and operational lives ranging from one week to more than ten years are employed to carry out these space activities.

These spacecraft are placed into orbit by a wide variety of launch vehicles. These launch vehicles, which may be either solid or liquid fueled, use multiple stages (some of which may themselves go into orbit) to place spacecraft into their desired orbit. Some spacecraft retain the stage used to perform their orbital insertion maneuver, and most spacecraft have some propulsive capability for attitude control and performing or-

---

BOX 1-1   Examples of Heavily Used Orbital Regions

Low Earth Orbit (LEO): A majority of the world's spacecraft operate in LEO because these orbits have characteristics that are advantageous for a wide array of missions. First, less energy (and thus a smaller launch vehicle) is required to launch a spacecraft into LEO than to put it into any higher orbit. Second, proximity to Earth allows remote sensing missions to receive higher resolution images. Finally, the Earth's magnetic field protects spacecraft in some LEOs from cosmic radiation and solar flares; this is of particular importance for human operations in space.

Sun-Synchronous Orbit: These LEOs precess in such a way that they do not experience changes in Sun angle due to the movement of the Earth around the Sun. This means that the lighting conditions for points on the Earth as the spacecraft passes overhead do not change over the course of a year—a useful feature for some remote sensing missions. Sun-synchronous orbits have inclinations greater than 90 degrees (the exact inclination varies with altitude). Although spacecraft can occupy Sun-synchronous orbits at most altitudes, for a number of reasons the altitudes near 900 and 1,500 km are the most widely used.

Geosynchronous Earth Orbit (GEO): GEOs are circular with orbital periods of approximately 1,436 minutes (about 24 hours), so spacecraft in them remain above roughly the same longitude on the Earth throughout their orbit. A special type of GEO is the geostationary Earth orbit, which has an inclination close to zero degrees. From the surface of the Earth, spacecraft in geostationary Earth orbits appear to be fixed in the sky. Communications with the spacecraft are thus simplified—both because the spacecraft is in view at all times and because ground antennas do not have to follow the spacecraft's movement. Inclined GEOs are also useful for some missions, although they require ground stations that are able to track a spacecraft's north-south as well as its apparent east-west movement.

---

bital corrections. These spacecraft are placed into orbits from which they can accomplish their particular mission effectively, resulting in a highly nonuniform distribution of spacecraft about the Earth. Box 1-1 lists three heavily used orbital regions and some of the reasons why they are used. (Additional information about these and other orbital regions is contained in the Glossary.) A few spacecraft each year are launched out of Earth orbit and into interplanetary space; the hazard to future space operations from these probes is utterly negligible.

The distribution of spacecraft around the Earth at the start of 1994 is displayed in Figure 1-1. This distribution is not static; as missions, technologies, and available launch vehicles change, the distribution of functional spacecraft also changes. For example, over the past three decades, the annual percentage of new space missions to orbits above LEO has been increasing; in 1993, High Earth Orbits (HEOs) were the final destination of 42 percent of successful launchings worldwide. Proposed future constellations of communications spacecraft in LEO may reverse this trend.

Figure 1-1 reveals a few characteristics of the current spacecraft population. Most spacecraft reside in LEO, but there are three significant concentrations in higher orbits. There is a concentration of spacecraft (performing Earth observation and communications missions) in GEO. A second concentration in and near circular semisynchronous orbits is made up of spacecraft from the U.S. Global Positioning System (GPS) and the Russian Global Navigation Satellite System. There is also a significant population of spacecraft in highly elliptical Molniya-type orbits, including Commonwealth of Independent States (CIS) early warning and communications constellations. (In this report, we will refer to pre-1992 space activities of the former USSR as "Soviet" and those of 1992 and later as either "Russian" or of the CIS, as appropriate.) In LEO, notable peaks exist around 1,400 to 1,500 km (due in part to a constellation of Russian communication spacecraft and debris from three breakups of Delta rocket bodies) and 900 to 1,000 km (due in part to Sun-synchronous remote sensing and navigation spacecraft and their associated debris).

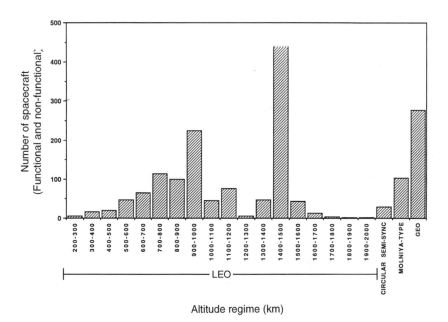

FIGURE 1-1 Spacecraft population in Earth orbit, 1994. SOURCE: Prepared by Kaman Sciences based in part on U.S. Space Command Satellite Catalog.

Most space activities involving humans occur below about 600 km; there are currently few spacecraft in these low orbits because atmospheric drag at these altitudes causes fairly rapid orbital decay.

TYPES OF ORBITAL DEBRIS

The more than 3,600 space missions since 1957 have left a legacy of thousands of large and perhaps tens of millions of medium-sized debris objects in near-Earth space. Unlike meteoroids, which pass through and leave the near-Earth area, artificial space debris orbit the Earth and may remain in orbit for long periods of time. Of the 23,000 space objects officially cataloged by the U.S. Space Surveillance Network (SSN) since the beginning of the space age, nearly one-third remain in orbit about the Earth; the majority of these are expected to stay in orbit for tens or hundreds of years. The increasing population of cataloged space objects is represented in Figure 1-2. It is imperative to note that this figure shows only the objects large enough to be repeatedly tracked by ground-based radar. The vast majority of debris is too small to be tracked and is not represented in the figure.

Objects in Earth orbit that are not functional spacecraft are consid-

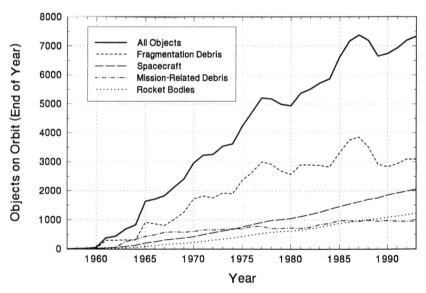

FIGURE 1-2 On-orbit cataloged population (corrected for delayed cataloging). SOURCE: Prepared by Kaman Sciences Corporation based in part on U.S. Space Command Satellite Catalog.

Box 1-2   Debris Size Conventions

This report uses three general size ranges to categorize debris.

Size Category	Approximate Diameter	Approximate Mass	Detectability	Probable Damage to Spacecraft
Large	>10 cm	>1 kg	May be catalogable in LEO	Probable loss of spacecraft and possible catastrophic breakup
Medium	1 mm-10 cm	1 mg-1 kg	Too small to catalog, too few for most in situ sampling	Ranges from surface degradation through component damage and possible loss of spacecraft capability
Small	<1 mm	<1 mg	Detectable by in situ sampling	Degradation of surfaces and possible damage to unprotected components

NOTE: Lines between categories cannot be sharply drawn. For example, Chapter 2 describes how the sizes of objects that are detectable through various means vary depending on sensor capability and the object's altitude, and Chapter 4 describes how damage caused by debris impact depends on the collision velocity and the particular configuration of the spacecraft being struck.

---

ered debris. Spacecraft that are passive, serving as platforms for laser ranging experiments, atmospheric density measurements, or calibration of space surveillance sensors are considered functional, as are spacecraft that are currently in a reserve or standby status and may be reactivated in the future to continue their mission. Each other type of object in Earth orbit may be classified as belonging to one of four types of debris: nonfunctional spacecraft, rocket bodies, mission-related debris, and fragmentation debris. Figure 1-3 indicates the relative numbers of cataloged functional spacecraft and debris as of March 1994. More than 90% of these cataloged space objects are of U.S. or Soviet/CIS origin, while the remainder belong to nearly 30 other countries or organizations.

Nonfunctional Spacecraft

Functional spacecraft represent only about one-fifth of the spacecraft population in Earth orbit—the large majority of orbiting spacecraft are

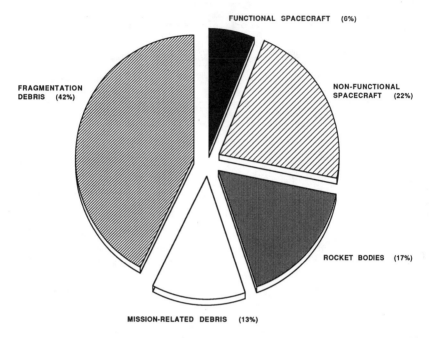

FUNCTIONAL SPACECRAFT (6%)

FRAGMENTATION
DEBRIS (42%)

NON-FUNCTIONAL
SPACECRAFT (22%)

ROCKET BODIES (17%)

MISSION-RELATED DEBRIS (13%)

FIGURE 1-3 Cataloged space objects by category, 1994. SOURCE: Prepared by Kaman Sciences Corporation based in part on U.S. Space Command Satellite Catalog.

nonfunctional. With very few exceptions, functional spacecraft that reach their end of life (EOL), through either termination or malfunction, are left in their former orbit or are transferred to slightly higher or lower altitudes (i.e., are reorbited). Typically, EOL reorbiting maneuvers are performed only by geosynchronous or semisynchronous spacecraft and by LEO spacecraft carrying nuclear materials. Historically, these EOL maneuvers have almost always resulted in longer orbital lifetimes. Only crewed vehicles and a few other spacecraft (e.g., reconnaissance or space station related) in very low orbits are normally returned to Earth at the conclusion of their missions.

BOX 1-3 Museum Piece in Orbit

The oldest nonfunctional spacecraft in orbit is the 1.5-kg Vanguard 1 spacecraft, which was launched by the United States on March 17, 1958, and ceased to function in 1964. Vanguard 1 now resides in an elliptical orbit of 650 km by 3,865 km. This orbit may not decay completely for another 200 years or more.

Rocket Bodies

The majority of functional spacecraft are accompanied into Earth orbit by one or more stages (or "rocket bodies") of the vehicles that launched them. Usually only one rocket body is left in orbit for missions to LEO, but the launch vehicle of a high-altitude spacecraft such as GOES (Geostationary Operational Environmental Satellite) may release up to three separate rocket bodies in different orbits along the way to its final destination. Relatively few spacecraft types (e.g., the U.S. National Oceanic and Atmospheric Administration and Defense Meteorological Satellite Program meteorological spacecraft and Soviet nuclear-powered ocean reconnaissance spacecraft) are designed to retain their orbital insertion stages and leave no independent rocket bodies. Figure 1-4 depicts the rocket bodies and other large debris placed into various orbits in the course of a Proton launch vehicle's delivery of a payload to GEO.

The presence of rocket bodies in orbit is of particular importance to the future evolution of the Earth's debris population due to their characteristically large dimensions and to the potentially explosive residual propellants and other energy sources they may contain. Although the largest stages, which are generally used to deliver spacecraft and any additional stages into LEO, usually reenter the atmosphere rapidly, smaller stages used to transfer spacecraft into higher orbits and insert

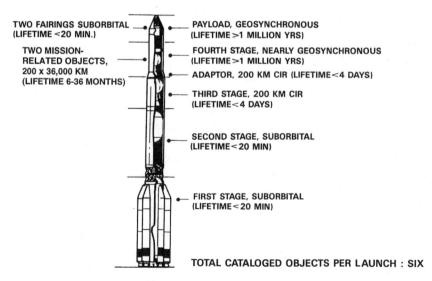

TWO FAIRINGS SUBORBITAL
(LIFETIME <20 MIN.)

TWO MISSION-
RELATED OBJECTS,
200 x 36,000 KM
(LIFETIME 6-36 MONTHS)

PAYLOAD, GEOSYNCHRONOUS
(LIFETIME >1 MILLION YRS)

FOURTH STAGE, NEARLY GEOSYNCHRONOUS
(LIFETIME >1 MILLION YRS)

ADAPTOR, 200 KM CIR (LIFETIME <4 DAYS)

THIRD STAGE, 200 KM CIR
(LIFETIME <4 DAYS)

SECOND STAGE, SUBORBITAL
(LIFETIME <20 MIN)

FIRST STAGE, SUBORBITAL
(LIFETIME <20 MIN)

TOTAL CATALOGED OBJECTS PER LAUNCH : SIX

FIGURE 1-4 Typical debris produced in a Proton launch to GEO. SOURCE: Teledyne Brown Engineering.

them into those orbits may remain in orbit for long periods of time. Many of these rocket bodies are in orbits that intersect those used by functional spacecraft.

Mission-Related Debris

Other space objects, referred to as mission-related debris, may be released as a result of a spacecraft's deployment, activation, and operation. Parts of explosive bolts, spring release mechanisms, or spin-up devices may be ejected during the staging and spacecraft separation process. Shortly after entering orbit, the spacecraft may release cords securing solar panels and other appendages or eject protective coverings from payload and attitude control sensors. The amount of debris released can be quite large; a detailed study of the debris released by one Russian launch mission revealed that 76 separate objects were released into orbit from either the launch vehicle or the spacecraft. Numerous debris may also be created during a spacecraft's active life. For example, during the first eight years of its operation, more than 200 pieces of mission-related debris linked with the Mir space station were cataloged. Although the occasional item accidentally dropped by a cosmonaut or astronaut may be newsworthy, the majority of this type of debris is intentionally dumped refuse. Since mission-related debris are often relatively small, only the larger items can be detected and cataloged by present-day ground-based surveillance networks.

Another type of mission-related debris comes from the operation of solid rocket motors normally used as final transfer stages, particularly on GEO missions. Current solid rocket fuel usually employs significant quantities of aluminum mixed with the propellant to dampen burn rate instabilities. However, during the burning process, large numbers of aluminum oxide (Al_2O_3) particles are formed and ejected through a wide range of flight path angles at velocities up to 4 km/s. These particles are generally believed to be no larger than 10 microns in diameter, but as many as 10^{20} may be created during the firing of a single solid rocket motor, depending on the distribution of sizes produced. While the orbital lifetimes of individual particles are relatively short, a considerable average population is suggested by examinations of impacts on exposed spacecraft surfaces. More than 25 solid rocket motor firings were conducted in orbit during 1993.

More recently, attention has been drawn to another side effect of solid rocket motors. Ground tests indicate that in addition to the large number of small particles, a smaller number of 1-cm or larger lumps of Al_2O_3 are also ejected during nominal burns. The only indication of the existence of such objects are data from ground tests carried out at

Marshall Space Flight Center, Alabama, and the Arnold Engineering and Development Center (Siebold et al., 1993). These medium-sized particles, which have lower characteristic ejection velocities and smaller area-to-mass ratios than the smaller particles, may also be longer-lived than the small particles and could pose a long-term hazard to other Earth-orbiting space objects.

Fragmentation Debris

Fragmentation debris—the single largest element of the cataloged Earth-orbiting space object population—consists of space objects created during breakups and the products of deterioration. Breakups are typically destructive events that generate numerous smaller objects with a wide range of initial velocities. Breakups may be accidental (e.g., due to a propulsion system malfunction) or the result of intentional actions (e.g., space weapons tests). They may be caused by internal explosions or by an unplanned or deliberate collision with another orbiting object.

Since 1961, more than 120 known breakups have resulted in approximately 8,100 cataloged items of fragmentation debris, more than 3,100 of which remain in orbit. Fragmentation debris thus currently makes up more than 40 percent of the U.S. space object catalog (and undoubtedly represents an even larger fraction of uncataloged objects). The most intensive breakup on record was the 1987 breakup of the Soviet Kosmos 1813, which generated approximately 850 fragments detectable from the Earth. The fragmentation debris released from a breakup will be ejected at a variety of initial velocities. As a result of their varying velocities, the fragments will spread out into a toroidal cloud that will eventually expand until it is bounded only by the limits of the maximum inclinations and altitudes of the debris. This process is illustrated in Figure 1-5. The rate at which the toroidal cloud evolves depends on both the original spacecraft's orbital characteristics and the velocity imparted to the fragments; in general, the greater the spread of the initial velocity of the fragments, the faster will the evolution occur.

In contrast, debris fragments that are the product of deterioration usually separate at low relative velocity from a spacecraft or rocket body that remains essentially intact. Products of deterioration large enough to be detected from Earth are occasionally seen—probably such items as thermal blankets, protective shields, or solar panels. Most such deterioration is believed to be the result of harsh environmental factors, such as atomic oxygen, radiation, and thermal cycling. During 1993 the still-functional COBE (Cosmic Background Explorer) spacecraft released at least 40 objects detectable from Earth—possibly debonded thermal blanket segments—in a nine-month period, perhaps as a result of thermal shock.

FIGURE 1-5 Evolution of
a debris cloud. SOURCE:
Kaman Sciences Corpora-
tion.

Phase 1

Phase 2

Phase 3

Another serious degradation problem involves the flaking of small
paint chips as a space object ages under the influence of solar radiation,
atomic oxygen, and other forces. Paint, which is used extensively on
both spacecraft and rocket bodies for thermal control reasons, can deteriorate severely in space, sometimes in a matter of only a few years. The
potential magnitude of the problem was not fully recognized until the
1983 flight of the STS-7 Space Shuttle mission, when an impact crater on

---

Box 1-4 Degradation Debris From LDEF

A variety of small and medium-sized debris is known to have been created by the degradation of surfaces on the LDEF spacecraft. Several multilayer insulation (MLI) blankets on the space-facing end came partially loose when the Kapton tape holding them to the spacecraft became brittle in the ultraviolet light exposure. Subsequent shrinking of the top face sheets of the MLI blankets stressed the embrittled Kapton tape and caused it to crack, partially releasing the MLI blankets (See et al., 1990; Adams et al., 1991). Several solar array specimens (each of which was approximately 5 cm by 10 cm) also came loose from LDEF. These specimens were mounted on Kapton substrates that were eroded by atomic oxygen exposure. (Whitaker and Young, 1991). The astronauts on board the shuttle visually identified (and filmed) one of the released solar array specimens as they approached LDEF during its retrieval mission. The films from this mission also show a cloud of small particles surrounding LDEF.

---

an orbiter window was apparently caused by a paint chip smaller than a millimeter in diameter. Subsequent analyses of spacecraft components returned from LEO have confirmed the presence of a large population of paint particles, even though the orbits of individual particles decay quite rapidly.

PERTURBATION FORCES AFFECTING SPACE OBJECTS

Once in orbit, debris is affected by perturbing forces that can alter its trajectory and even remove it completely from orbit. Other than the gravitational attraction of the Earth, the primary forces acting on a space object in lower orbits (below about 800 km) are atmospheric drag and gravitational perturbations from the Earth. These gravitational perturbations, however, although affecting some orbital parameters, do not generally strongly affect orbital lifetime. For space objects in higher orbits, solar and lunar gravitational influences become more important factors. Small debris can also be affected by solar radiation pressure, plasma drag, and electrodynamic forces, although the effects of plasma drag and electrodynamic forces are typically dwarfed by the effects of solar radiation pressure.

The rate at which a space object loses altitude is a function of its mass, its average cross-sectional area impinging on the atmosphere, and the atmospheric density. Although the Earth's atmosphere technically extends to great heights, its retarding effect on space objects falls off rapidly with increasing altitude. Atmospheric density at a given altitude, however, is not constant and can vary significantly (particularly at less than 1,000 km) due to atmospheric heating associated with the 11-year solar cycle. This natural phenomenon has the effect of accelerating the orbital decay of debris during periods of solar maximum (increased sunspot activity and energy emissions). During the last two peaks in the solar cycle, the total cataloged space object population actually declined, because the rate of orbital decay exceeded the rate of space object generation via new launches and fragmentations.

Figure 1-6, which displays the predicted orbital lifetimes for a number of different objects in circular LEOs at different periods in the solar cycle, illustrates the importance of cross-sectional-area-to-mass ratio, altitude, and solar activity in determining orbital lifetimes in LEO. First, objects with low ratios of cross-sectional area to mass decay much more slowly than objects with high area-to-mass ratios. Second, objects at low altitude experience more rapid orbital decay than objects at high altitude. Finally, objects decay much more rapidly during periods of solar maximum than during the solar minimum.

Solar-lunar gravitational perturbations primarily affect the orbital lifetimes of space objects in highly elliptical orbits (e.g., Molniya-class or Geostationary Transfer Orbits [GTO]). Depending on the alignments of the space object's orbital plane with the Moon and the Sun, these forces can either accelerate or decelerate the orbital decay process substantially. For example, a GTO rocket body could reenter the Earth's atmosphere within a few months or remain in orbit for more than a century, depending on the position of the Sun and the Moon at the time of its injection into transfer orbit. GEO missions launched by the CIS routinely take advantage of these forces to limit the lifetime of GTO debris to less than three years, with many objects decaying in less than six months.

Solar radiation pressure normally induces a noticeable effect on a space object's orbit if that object possesses a large area-to-mass ratio. These effects can lead to an increase in the eccentricity of the orbit, which in turn leads to more rapid decay if the resulting lower perigee exposes the space object to significantly greater atmospheric density levels. Insulation materials and inflatable space objects are often strongly affected by solar radiation pressure. Debris from the ruptured Pageos balloon, for example, exhibited strong orbital perturbations due to solar radiation pressure, as has some debris from more conventional fragmentations.

The combination of all of these forces has caused approximately 16,000 cataloged objects to reenter the atmosphere since the beginning of the space era. In recent years, an average of two to three space objects large enough to be cataloged (as well as numerous smaller debris particles) reenter the Earth's atmosphere each day. Over the course of a year, this amounts to hundreds of metric tons of material. This material is

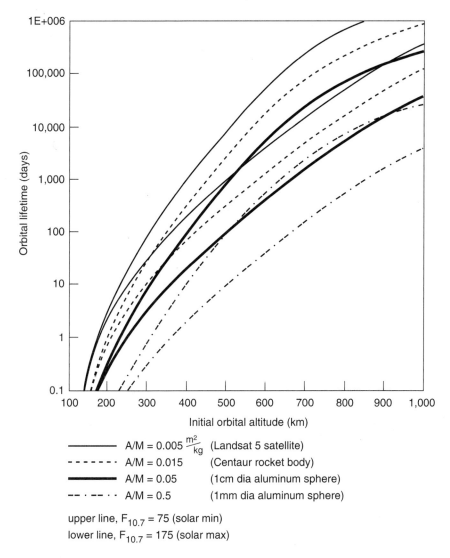

FIGURE 1-6 Orbital decay time versus altitude. SOURCE: National Aeronautics and Space Administration.

composed primarily of large objects that were launched into low orbits (most of the mass is in the form of large multiton rocket bodies) and small objects with high cross-sectional-area-to-mass ratios. Seldom do any large objects initially placed into orbits higher than 600 km reenter the atmosphere.

FINDING

Finding 1: Orbital debris travels in a variety of orbits and is affected by various perturbation forces, including the effects of the Earth's atmosphere, gravitational perturbation effects, and solar radiation pressure. As orbital altitude increases, the effect of the atmosphere in accelerating orbital decay becomes small, and typical large objects in orbits higher than approximately 600 km can remain in orbit for tens, hundreds, or even thousands of years.

REFERENCES

Adams, J.H., L.P. Beahm, and A.J. Tylka. 1991. Preliminary results from the heavy ions in space experiment. P. 377 in NASA Conference Publication 3134, LDEF—69 Months in Space: Proceedings of the First Post-Retrieval Symposium, Kissimmee, Florida, June 2–8. A.S. Levine, ed. Hampton, Virginia: NASA Langley Research Center.

See, T., M. Allbrooks, D. Atkinson, C. Simon, and M. Zolensky. 1990. Meteoroid and Debris Impact Features Documented on the Long Duration Exposure Facility: A Preliminary Report. NASA JSC #24608. Houston, Texas: NASA Johnson Space Center.

Siebold, K.H., M.J. Matney, G.W. Ojakangas, and B.J. Anderson. 1993. Risk analysis of 1-2 cm debris population for solid rocket motors and mitigation possibilities for geotransfer orbits. Pp. 349–351 in Proceedings of the First European Conference on Space Debris, Darmstadt, Germany, April 5–7. Darmstadt: European Space Operations Center.

U.S. Space Command. 1994. U.S. Space Command Satellite Catalog. Cheyenne Mountain Air Force Base, Colorado: U.S. Space Command.

Whitaker, A.F., and L.E. Young. 1991. An overview of the first results on the Solar Array Materials Passive LDEF Experiment (SAMPLE), A0171. P. 1241 in NASA Conference Publication 3134, LDEF-69 Months in Space: Proceedings of the First Post-Retrieval Symposium, Kissimmee, Florida, June 2–8. A.S. Levine, ed. Hampton, Virginia: NASA Langley Research Center.

2

Methods for Characterization

Information about the orbital debris environment is needed to determine the current and future hazard to space operations from debris. Unfortunately, the debris environment is difficult to characterize accurately. Only the largest debris objects can be repeatedly tracked by ground-based sensors; detection and tracking of the numerous smaller pieces of debris is much more difficult. A variety of measurement techniques have been developed, however, that enable statistical estimates to be made of the number and characteristics of some size ranges of smaller debris items in some orbits. These estimates rely on scientific and engineering models of population characteristics. More complex models are used to estimate the characteristics of the future debris population.

TRACKING AND CATALOGING ORBITAL DEBRIS

Current Capabilities

A small percentage of debris in orbit is tracked and cataloged. The orbital parameters (e.g., period, inclination, apogee, and perigee) of these objects are entered into a catalog, generally along with information on the object's origin—only objects with known origins are entered into the catalog—and its radar or optical cross section. These data can then be used for such purposes as predicting potential collisions and recognizing space object breakups. Cataloging space objects requires an expensive network of sensors capable of observing objects periodically to determine any changes in their orbital elements and of continually performing orbit determination computations.

Currently, only two systems in the world have the necessary network of ground-based sensors and computational capability to carry out this task. One, the Space Surveillance Network (SSN), is operated by the United States under the control of the U.S. Space Command; the other, the Space Surveillance System (SSS), is operated by the Russian military (see Box 2-1). The primary purpose of each system is to detect objects that present a military threat; thus, although each is capable of detecting certain types of debris, neither system is optimized to perform the task of maintaining a debris catalog.

BOX 2-1 The Russian and U.S. Space Surveillance Systems

Figure 2-1 displays the location of the sensors of the Russian and U.S. space surveillance systems.

The Russian Space Surveillance System (SSS) has a primary data acquisition system that includes 10 radars (operating in either UHF [ultrahigh frequency], VHF [very high frequency], or C-band) and 12 optical and electro-optical facilities. The radars are used to track objects in lower orbits; the optical and electro-optical facilities are used only for tracking objects in high orbits. Additional sensors may be used occasionally for important tasks and experiments. The lack of a worldwide network of sensors results in some major breaks in observation and some zones in which space objects cannot be observed.

Data from the sensors (approximately 50,000 measurements per day) are transmitted to the Russian Space Surveillance Center, where they are processed, and the space object catalog is updated and replenished. The Russian Space Surveillance Center also identifies detected objects, updates space object orbital elements and calculates orbital element sets for new observations, plans future observations, determines orbital lifetimes, and provides information to other space programs (Batyr et al., 1993; Batyr et al., 1994).

The U.S. Space Surveillance Network (SSN) consists of more than 20 radar and optical sensors, most of which are not dedicated to space surveillance and are tasked on an "as-needed" basis. In general, radars are used to track objects in low-altitude orbits and optical sensors are used for high-altitude detection; some radars, however, are deep-space sensors capable of detecting objects in GEO. Although many of the SSN's sensors are located within the continental U.S., others are spread out longitudinally.

Data from the network are fed to the U.S. Space Control Center, which processes the data and maintains a catalog of space objects. Orbital element sets are transmitted back to the sensors to allow them to continue tracking detected objects and are also made available to selected satellite operators and space system users. The Space Control Center also processes space object breakup data and performs collision warning for some space activities, such as launches and U.S. Space Shuttle operations in orbit.

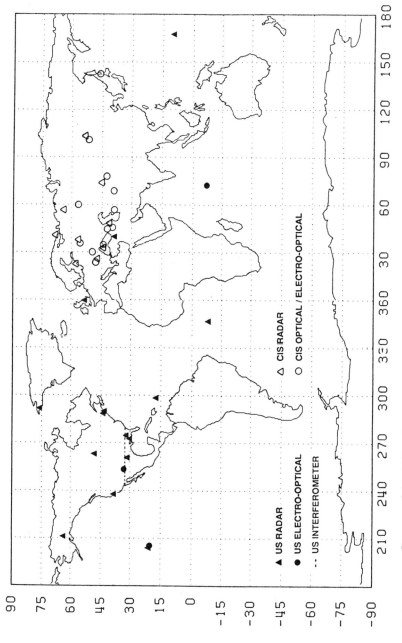

FIGURE 2-1 Sensors of the SSS and SSN. SOURCE: Kaman Sciences Corporation.

The ease with which a particular space object can be tracked depends on its optical or radar cross section (RCS) as well as its orbital parameters. In general, objects with larger optical or RCSs are more easily detectable than objects with smaller cross sections. Both the optical and the radar cross sections of particular space objects can vary greatly—which is not surprising for a collection of irregular-shaped objects. Uncertainty in the relation of RCS to actual size means that the smallest objects that these systems are able to catalog is uncertain, but since few objects in the SSS or SSN catalogs have an RCS of less than about 0.01 square meter, the commonly reported minimum trackable size has been 10 cm in diameter.

Recent radar range calibration of fragments produced in the laboratory, combined with measurements by short-wavelength radar and by ground telescopes (Henize and Stanley, 1990), have provided additional insight into the limiting size of the objects maintained in the catalog. These data indicate that for LEO orbital inclinations above about 30 degrees, the U.S. catalog contains some objects as small as 10 cm but is not complete at this size range. The catalog for LEO objects with inclinations greater than 30 degrees, however, is estimated to be 90 to 99 percent complete for objects larger than 20 cm.

As orbital altitude increases, the minimum size of debris that can be detected by ground-based sensors increases. However, this does not mean that the minimum-sized object that can be *cataloged* increases steadily with altitude. The opportunity for repeated observations and the predictability of an object's position in orbit also increase with altitude, making the maintenance of the orbital elements of a high-altitude detected object easier. Consequently, for altitudes below about 2,000 km, there is no simple statement of the limiting size of the catalog, other than that it is in the 10- to 30-cm range. However, radar detection sensitivity rapidly decreases with increasing altitude, and by 5,000 km the smallest

---

BOX 2-2   Comparison of the SSN and SSS Catalogs

The U.S. and Russian space object catalogs are in general agreement for LEO objects greater than 50 cm in diameter. For space objects with diameters between 10 and 50 cm, the U.S. catalog is more complete. Above LEO, both catalogs generally maintain the orbital elements only of spacecraft and rocket bodies greater than 1 meter in diameter. Due to the lack of a worldwide network of sensors, the Russian space object catalog does not include objects in a significant portion of GEO and can only periodically track objects in highly eccentric, low-inclination orbits.

objects detectable by radar are about 1 meter in diameter. Above 5,000 km, optical telescopes become the primary sensors; these have the sensitivity to track meter-sized objects in GEO—though this does not mean that all meter-sized objects in GEO are cataloged.

Current space surveillance systems have difficulties in cataloging some space objects in highly elliptical orbits and low-inclination orbits. Objects in highly elliptical orbits are difficult to detect because they spend a large fraction of their time at very high altitudes, while objects in low-inclination orbits are more difficult to detect because of the relative lack of sensors (in either network) at low latitudes. Recent experiments by the U.S. Space Command confirmed the SSN's difficulty in cataloging space objects in low-inclination and high-eccentricity orbits (Pearce et al., in press; Clark and Pearce, 1993). It should be emphasized that these peculiarities do not represent deficiencies in the way the networks perform their normal mission of maintaining a catalog for military reasons, but rather reflect the fact that they were not designed to characterize the space debris population.

Improving Tracking and Cataloging Capabilities

International cooperation might present an opportunity to make some improvement in the catalogs without significant expenditure. The SSS and the SSN both routinely track objects not found in the other's catalog, so sharing catalog data will improve the completeness of both catalogs. (It is not at all clear, however, that sharing catalog data would increase the accuracy with which the orbital parameters of cataloged objects are known.) Since both the SSS and the SSN have similar limitations, it is already clear that information sharing between the two systems would not significantly increase the size of the catalog or improve detection of medium-sized debris. There is also a potential obstacle to such collaboration in that there are legitimate security reasons for not sharing all data received from national surveillance networks; this may not be a major issue because both networks are capable of editing data before sharing them.

One factor that limits the ability of most space surveillance sensors to detect smaller debris is that they were not designed to detect small objects. Most space surveillance radars operate in the UHF and VHF ranges; debris smaller than about 10 cm in diameter are in the Rayleigh scattering region for these frequencies and are thus not easily detected, and the record of their orbital elements is not easily kept current. A National Aeronautics and Space Administration (NASA) study on the possible protection of the Space Station against debris concluded that 10 cm was an inherent limit for the current sensors of the SSN and that these sensors could not easily be modified to improve sensitivity (NASA, 1990). The Russian SSS is currently working to increase its capability to observe small objects with existing sensors, focusing research on lowering the sensitivity thresholds of its current radars and on developing new methods to acquire weak signals using narrow-angle and narrow-beam sensors and making full use of existing data regarding the space object's motion. While this research may allow the SSS to track somewhat smaller debris, radars operating at much shorter wavelengths (e.g., 3 cm wavelength to detect 1 cm diameter objects in LEO) will ultimately be required to detect debris significantly smaller than 10 cm in diameter.

Increasing the accuracy of predictions of the future location of objects in LEO is another means of improving tracking and cataloging capabilities. Such improvement is a necessary requirement for the development of an effective collision warning capability in LEO; increased accuracy is required to keep the number of false alarms for such a system low, since moving spacecraft is a task not undertaken lightly. (Collision warning schemes are discussed in some detail in Chapter 7.) Currently, uncertainty in the future location of objects due to atmospheric drag is the major limitation on catalog accuracy in LEO. This unavoidable uncertainty is due to variability in the density of the upper atmosphere and uncertainty about objects' orbital attitude (and thus the cross-sectional area they present to the atmosphere) and normally dwarfs inaccuracies caused by observation errors and errors in propagation theory. As is shown in Figure 2-2, atmospheric drag retardation along the orbital track of medium to large space objects in 300- to 600-km-altitude orbits can range up to hundreds of kilometers per day.

The most optimistic estimate of the accuracy with which atmospheric drag can be determined is ±15 percent; consequently a prediction error (which cannot be calibrated) of several kilometers per day is typically accumulated. Keeping the number of false alarms for a LEO collision warning system at a tolerable level thus requires frequent reobservations of debris objects. (Collision warning systems for objects in regions where atmospheric drag is less critical would not have this limitation; in GEO, for example, errors in estimates of objects' initial positions would be responsible for the majority of false alarms.) Improvements in propagation accuracy could be achieved by positioning sensors to minimize the required propagation time and by improving understanding of upper-atmospheric density fluctuations.

Improving the ability to track and catalog objects in orbits above LEO is basically a matter of improving the sensors (both radar and optical) used to detect high-altitude objects and acquiring enough data from these sensors to determine the orbital parameters of the objects they detect. Detecting objects that are less bright (because they either are smaller,

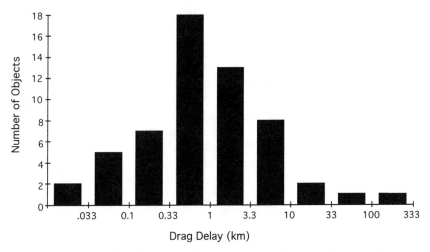

FIGURE 2-2 One day along-track drag retardation for a random sample of cataloged objects at 300–600 km. SOURCE: U.S. Naval Space Command Satellite Catalog.

are further away, or have a lower albedo) might be accomplished with either larger-aperture telescopes or telescopes equipped with charge-coupled devices (CCDs). Siting debris-detecting sensors at low latitudes could allow a broad variety of objects, including those in low-inclination orbits, to be detected. Finally, increasing the number of sensors available to detect debris would allow for better tracking of cataloged objects and for more searches for uncataloged objects.

BOX 2-3 Detecting Debris with CCDs

Charge-coupled devices can be used in optical sensors to convert incoming light directly to electric charges; the magnitude of the output signal is proportional to the light intensity. CCDs have not yet outperformed conventional sensors for detection of objects in LEO because the rapid movement of LEO objects requires that the signal be integrated, which in turn requires an assumption of direction of motion, severely limiting the detection rate. CCDs are improving, however, and are already outperforming non-CCD sensors for observation of high-altitude debris (which does not move as rapidly across the field of view). Upgrading the SSN's GEODSS (ground-based electro-optical deep-space sensors) to use CCDs has been considered.

SAMPLING ORBITAL DEBRIS

Since it is currently impossible to track all debris in orbit, measuring and characterizing the uncataloged debris population must be carried out by sampling the debris flux at particular locations and times and using the data as a basis for estimating the characteristics of the general population. The orbital debris flux can be sampled either directly (with spacecraft surfaces that are later returned to Earth for examination) or remotely (using ground- or space-based radars or optical telescopes that record debris as it passes through their fields of view). Although sampling—combined with predictive models—can be used to provide important clues to the nature of debris populations that are not included in the catalog, it is important that the limitations of the technique, including any sampling biases, be taken into account. For example, rather than portraying the steady-state small debris population in LEO, in situ measurements of small debris particles acquired by examining returned spacecraft surfaces portray only the average debris flux along a particular orbit during a particular time frame.

Remote Sampling from Earth

Optical Sensors

At first glance, the use of ground-based telescopes to sample the debris population seems like a promising technique. Such sampling is usually carried out by pointing the telescope in a fixed direction and counting objects as they pass through its field of view. A 1-meter diameter telescope in darkness can theoretically detect a sunlit metal sphere 1 cm in diameter at 900-km distance. If this were all there was to the problem, data from optical sensors could be used to estimate the population of objects larger than 1 cm in diameter in orbits up to 900 km.

Unfortunately, most debris fragments reflect much less light than a metal sphere; typically only about 10 percent of the light is reflected. In addition, objects in LEO have angular velocities of at least 0.5 degree per second when viewed from the ground, which further increases the difficulty of optical detection. Finally, there can be difficulty in discriminating between debris and the luminosity caused by meteors interacting with the atmosphere. Theoretically, this last problem can be solved completely by using two telescopes and determining the object's altitude with the measured parallax, or solved partially by using the object's angular velocity to approximate its altitude.

Despite these drawbacks, ground-based telescopes engaged in sampling have provided some valuable information on the LEO population of debris around 10 cm in diameter. Tests to detect uncataloged debris in

LEO with ground-based telescopes have been carried out by NASA (in cooperation with the Massachusetts Institute of Technology's Lincoln Laboratory and the U.S. Space Command) since 1983. These tests used electro-optical telescopes of approximately one 1-m diameter and, as mentioned earlier, aided in the determination of the approximate size ranges of debris contained in the SSN catalog. Although the exact size of debris detectable by these telescopes is not certain since they measure pieces of debris with a variety of unknown reflectivities, the average minimum object size detectable is slightly smaller than 10 cm (Kessler, 1993).

Ground-based telescopes also can be used to sample the space debris population above LEO. The limited efforts to sample the HEO population to date include surveys of GEO by the Russian Academy of Sciences and NASA, and surveys of GTO performed by ground-based electro-optical deep-space sensors sites. Tests to observe objects in the geostationary orbit with ground-based optical sensors have detected uncataloged objects, but there have been no comprehensive surveys of the geostationary ring and the size of its uncataloged population remains unclear.

Many of the features suggested earlier for improving the tracking and cataloging of high-altitude debris using optical sensors (e.g., larger apertures, low-inclination sites, or the use of CCDs) would also be useful for sampling the debris population. One additional feature particularly useful for sampling is a wide field of view, which gives an optical sensor the ability to sample large areas and thus gather more data. This is very useful in optical sensing, where the need for good lighting conditions can severely limit the hours a telescope can be used to look for objects in Earth orbit.

NASA is beginning to use a 3-m diameter "liquid-mirror" telescope to sample the debris population. Large liquid mirrors can be constructed relatively inexpensively because they use mercury, spun to keep it in the necessary parabolic shape, to form their reflecting surface. Such telescopes are constrained to always point vertically; although this makes some types of observation difficult, it does not hamper debris sampling. NASA finished construction of its first liquid mirror, which will operate within the United States, in 1994 and has already obtained stellar images from the telescope's temporary site in Houston. A second liquid-mirror telescope to be sited near the equator is planned. These telescopes should be able to regularly detect debris down to about 2-cm diameter at altitudes up to 500 km.

Radar Sampling

Short-wavelength ground-based radars also have been used effectively to sample the medium-sized debris population in LEO. Radars sample debris in a "beam park" mode (similar to the sampling technique used by ground telescopes), in which the radar stares in a fixed direction (preferably vertically to maximize sensitivity) and debris is counted as it passes through the radar's field of view. Since 1987, significant amounts of sampling data have been obtained by using the Arecibo, Goldstone, and Haystack radars. In addition, the longer-wavelength FGAN and MU radars have demonstrated the ability to sample the medium and large debris population, respectively (Mehrholz, 1993; Sato et al., 1992).

In 1989, the Arecibo Observatory's high-power 10-cm-wavelength radar and the Goldstone Deep Space Communications Complex's 3-cm-wavelength radar were used (with the assistance of other radars) to obtain orbital debris data. Neither radar was designed to track debris, but both were expected to detect small debris if it existed. In 18 hours of operation, the Arecibo experiment detected nearly 100 objects larger than an estimated 0.5 cm in diameter (Thompson et al., 1992). In 48 hours of observation, the Goldstone radar detected about 150 objects larger than approximately 0.2 cm in diameter (Goldstein and Randolph, 1990). Because little effort was made either to accurately define the collection area of these radars or to properly calibrate them, these data have fairly large uncertainties. Even so, these experiments demonstrated that data could be obtained in a beam park mode and that there was a large population of smaller debris to be detected.

Since 1990, more than 2,400 hours of data have been collected and analyzed from the Haystack radar (Stansbery et al., 1994). This 3-cm-wavelength radar situated at 42°N latitude can be pointed either vertically or south, 10 or 20 degrees above the horizon. In the vertical mode, maximum sensitivity is obtained, but detection in LEO is limited to orbits with inclinations greater than 42 degrees. When the radar is pointed south, sensitivity is poorer, but LEO objects with inclinations as low as 25 degrees can be detected. The complete data set from the Haystack observations contains information on the size, altitude, range rate (the rate of change in the distance from the object to the radar), and direction of motion of debris at altitudes up to 1,500 km. The data on the direction of motion can be used to determine an object's orbital inclination with a typical uncertainty of about 5 degrees (though uncertainty can be much higher for objects that are barely detectable). The range rate data can be used to determine orbital eccentricity when pointed vertically and inclination when pointed near the horizon. In the vertically pointing mode, the smallest objects detected range from about 0.3 cm at 350 km to 0.7 cm at 1,400 km. In the south-pointing mode the smallest objects detectable are larger—typically about 1 cm. Haystack transmits right circularly polarized radio waves and receives both left and right circularly polarized waves. The polarization of the reflection can be used to infer the general shape of the objects detected.

Calibration of the data acquired by using sampling radars can be achieved by a number of techniques. These include radar range measurements of fragments of known size, shape, and mass, and the use of orbital calibration spheres. The Haystack radar used both of these techniques. In this case, the range measurements indicated that irregular fragments reflected similarly to spheres but a broad distribution of possible signal returns must be considered in interpreting the data. Existing calibration spheres, as well as the Orbital Debris Radar Calibration Spheres (ODERACS), were also used in calibration.

Future efforts to sample the debris population with ground-based radars may be the most effective means to collect data on medium-sized debris in LEO. Improvements in this capability can be achieved by (1) performing more debris sampling with existing radars; (2) siting new radars so they can detect low-inclination debris populations; and (3) using high-powered, short-wavelength radars to detect smaller debris.

Increasing the amount of time that radars spend sampling debris is basically a problem of allocating the resources needed to carry out additional searches. Continued sampling efforts with existing radars can increase statistical confidence in existing data and, over time, could provide information on changes in the debris population. However, the Haystack, Goldstone, and Arecibo radars, which were not designed to detect debris, have other users preventing them from being used full-time for debris detection and are expensive to operate. For these reasons, the Haystack Auxiliary Radar (HAX) was recently built specifically to detect debris. HAX, which is located near the Haystack radar, will not be as sensitive as Haystack, but its slightly larger field of view and lower

---

BOX 2-4   The ODERACS Experiment

The ODERACS experiment was launched from the U.S. Space Shuttle in March of 1994 and provided calibration for a number of Earth-based radar and optical sensors. In this experiment, six aluminum spheres (two 5 cm in diameter, two 10 cm in diameter, and two 15 cm in diameter) were released into LEO. One sphere of each type had a polished surface whereas the other had a rough surface. This experiment demonstrated the validity of sampling debris with a radar and helped calibrate both radar and optical sensors. A similar future experiment will release three spheres and three dipoles to further calibrate the sensors. The dipoles are intended to calibrate polarization measurements, which are important for determining debris shape.

operating costs will allow data on low-altitude, medium-sized debris to be collected more rapidly and at a lower total expense. Data collection from this sensor began in 1994.

Locating a similar short-wavelength radar near the equator could further improve radar debris sampling over the capability of the HAX. Such a sensor could sample the entire LEO population, rather than just those objects with higher-inclination orbits. Comparisons between the populations detected by HAX and an equatorial sensor could also help illuminate the distribution of uncataloged debris by inclination.

Higher-power and shorter-wavelength radars might also improve sampling capabilities, although there are limits to such a strategy. A high-power, 1-cm-wavelength radar with a large antenna, for example, would be capable of acquiring data on debris as small as 0.1 cm in LEO, but the construction and operation of such radars could be very expensive. It is not feasible to detect even smaller debris from Earth by further decreasing the radar wavelength because the Earth's atmosphere absorbs radar signals with wavelengths smaller than about 1 cm.

Remote Sampling from Orbit

Remote sampling of debris from orbit could provide data on debris that are very difficult to detect from Earth, but there are a number of difficulties, both technical and economic, in such an approach. The major advantages of space-based sensors are that they can be much closer to orbital debris than ground-based sensors and that they do not have to peer through the Earth's atmosphere to see the debris. Their major disadvantage is the difficulty and cost of developing and deploying the powerful sensors typically required to detect these relatively small, fast-moving, and often dark objects.

Although no space-based remote sensor has yet been dedicated to debris observation, debris has been detected in the course of space-based astrophysical observations. The Infrared Astronomical Satellite, which was designed to image deep-space infrared sources, detected possible debris objects down to 1 mm in size at distances of up to 1,000 km, but difficulties in calibrating the sensors make extrapolating these results to the general population very problematic (de Jonge, 1993). The White Light Coronagraph on Skylab and the Hubble space telescope are also likely to have detected debris, but their data have not been analyzed for this purpose.

Many additional space-based remote sensors to detect debris have been studied and proposed since the late 1970s (Kessler and Cour-Palais, 1978; Neste et al., 1982). Among these are proposals by Russian experts, who proposed using space-based optoelectronic sensors to detect debris

---

BOX 2-5   Space-Based Remote Sensing of Debris: An Example

In late 1987, NASA began work on a proposal to detect 1-mm to 1-cm debris in LEO with a space-based electro-optical sensor. The project was called "Quicksat" because, to obtain a free rocket launch, the instrument had to be constructed rapidly. To obtain an expected detection rate (in a 500-km, sun-synchronous orbit) of about 200 objects per year, the initial design incorporated two 40-cm-diameter, F/1.0 telescopes using 500 by 500 CCD pixel arrays and separated by 1 meter to obtain usable parallax data. However, without on-board data processing, data transmission for this design was much too high. Consequently, fewer and larger pixels were required; however, this significantly increased the separation distance required for the two sensors to make parallax measurements. By optimizing the size, position and readout rate of the pixels, Quicksat was redesigned into two 25-cm-diameter, F/1.0 optics with 9 by 16 pixel arrays, decreasing the spacecraft mass by 200 kg and increasing the expected detection rate by a factor of 10. However, the cost of building and integrating the Quicksat satellite was estimated at more than $100 million; by 1988, the free rocket was unavailable and the project was canceled.

(Utkin et al., 1993); German experts, who proposed flying optical sensors on a space station or the U.S. Space Shuttle (Bendisch et al., 1993); and U.S. experts. In the United States, NASA (working with Ball Corporation) has done extensive work on an infrared system for debris detection and collision warning for the Space Station and has proposed a spacecraft (see Box 2-5) with two small telescopes capable of monitoring the 1-mm and larger environment (Portree and Loftus, 1993). In addition, Kaman Sciences Corporation has proposed an optical (visible and infrared) debris detection and characterization system for use on the U.S. Space Shuttle, and the U.S. Department of Energy's National Laboratories (specifically Sandia, Los Alamos, and Livermore), along with several companies, have each proposed various sensors (including radar, infrared, optical, and LIDAR) as potential space-based debris detection sensors. In early 1995, the U.S. Department of Defense plans to launch its MSX (Midcourse Space Experiment) spacecraft, which will use ultraviolet, infrared, and visible light sensors developed for other purposes to search for uncataloged debris.

Both passive or active remote sensors in orbit could theoretically be used to detect debris. Passive sensors (such as telescopes) detect objects by using existing illumination (such as reflected sunlight or the infrared radiation emitted by heated objects). Active sensors (such as LIDAR or radar) illuminate an object and detect the reflected illumination. One advantage of active sensors is that a single active sensor can accurately determine an object's range and the rate at which the range changes. In contrast, two passive sensors are needed to determine the range of an object, and even in pairs, current passive sensors would have difficulty in accurately determining the rate at which the range changes. This does not rule out the use of passive sensors for debris detection, however, because range and range rate determinations, while helpful in understanding the data collected, are not essential to obtaining useful data. Passive sensors also require less power than active sensors and have a smaller loss of sensitivity with range, so they are typically less expensive and will detect more objects than active sensors.

Since larger objects are easier to detect, the collection area of orbital remote sensors increases with increasing debris size. Although theoretically this should allow these sensors to detect a statistically meaningful sample over a large range of debris sizes, there are practical limitations, particularly in LEO. For example, small debris can be detected only so close to the detector that (in LEO) the expected high velocity at which the objects will pass across the field of view requires rapid readout rates, which increases noise for certain optical systems and increases the required data transmission rates for any system (unless on-board processing is used). For these reasons, it is currently difficult to observe small debris in LEO with space-based remote sensors. In contrast, debris larger than 1 cm can be detected at considerable distances, so the problems caused by movement across the sensor's field of view are much less severe. However, in the lower regions of LEO, it is more practical to detect debris in this size range from the surface of the Earth (because of the larger sensors available on the ground). Ground-based sensors may even be superior for the task of detecting LEO debris significantly smaller than 1 cm in diameter; the largest available ground-based radar used to detect debris (the Goldstone radar) can detect debris as small as 2 mm in diameter. Consequently, space-based remote sensors are likely to add significantly to our knowledge of the LEO debris environment only at higher LEO altitudes and in latitude bands that are not adequately characterized by ground-based sensors.

Space-based remote sensors offer more promise in HEOs, particularly GEO. At high altitudes, space-based sensors would be much closer to the debris being detected than Earth-based sensors and would have to detect only objects smaller than about a meter in diameter to improve on current measurements. In addition, objects in GEO would move relatively slowly across a sensor's field of view, greatly simplifying detection. However, either multiple sensors or sensors able to move along the geostationary ring would be needed to observe objects throughout GEO.

Impact Sampling

Currently (and for the foreseeable future), space debris particles smaller than a few millimeters in diameter cannot be detected by using Earth-based measurement techniques; effective remote sensing of such particles from orbit may also be infeasible. In situ impact techniques, however, can be used to sample this population effectively, characterizing particle sizes and materials as well as orbital distributions and dynamics (although such characterizations can be extremely difficult). Such measurements can be performed either passively, by exposing surfaces in orbit and then returning them to Earth for examination, or actively, by using any of a number of techniques ranging from impact detection with simple semiconductor-based sensors to chemical composition analysis of impacting particles with complex sensors.

There are a number of limitations on all current approaches to in situ debris sampling. First, impact detectors can sample only debris that intersects the orbit in which they are traveling. Second, the extent of the environment measured (in terms of the particle sizes that can be expected to impact the detector) and the statistical validity of the data are both dependent on the detector's total exposed area and the exposure time. Third, some measure of the impacting particle's velocity vector or its composition is needed to identify specific sources of impacting particles (i.e., whether they are meteoroids or orbital debris), and knowledge of the particle's velocity vector is also necessary to determine its pre-impact orbital parameters. Such knowledge is, however, difficult to acquire.

Passive Measurements

Passive in situ measurements of the debris environment are made by exposing samples of materials to the space environment and then returning them to Earth. Once on the ground, craters and perforations in the materials are measured and the diameters and impact velocities of the particles that caused this damage are estimated. Such data have been collected from the Apollo capsule windows, from Skylab exposure experiments, from U.S. Space Shuttle windows, from materials returned from the Solar Maximum Mission spacecraft, from the Salyut and Mir space stations, from the Palapa and Westar spacecraft, from LDEF, and from the European Retrievable Carrier (EURECA). Most recently, materials returned from the repair of the Hubble space telescope were made available for assessment.

Although these represent a significant number of surfaces over an extended period of time, the value of the data gathered in these experiments is limited. First, all of these data were acquired at altitudes between about 300 and 600 km; consequently, little is known about how the small debris population varies with altitude. Second, few of the surfaces returned to Earth for analysis have been adequately calibrated against one another; this makes analysis of variations in the impacting population over time difficult. In addition, passive sensors provide only integrated time-exposure data rather than time-dependent data, so little can be determined about the effects of solar activity on the small debris population or the existence and location of debris "swarms." Finally, because the majority of returned surfaces were not designed for debris testing, it is often difficult to distinguish between the impacts of orbital debris and micrometeoroids.

The applicability and validity of the damage scaling laws used to interpret the data from passively exposed detector materials are also an issue. Historically, the damage scaling laws used to estimate impactor characteristics from a surface crater or perforation have been derived empirically or are semiempirical. Different sets of scaling laws for interpretation have been applied to every set of impact data returned to date, and multiple different scaling laws (e.g., McDonnell and Sullivan, 1992; NASA, 1970) were used for LDEF and EURECA. Interpretations of impactor size by these different scaling laws vary by up to about a factor of three for typical "theoretical debris parameters" (e.g., spherical aluminum projectiles impacting an aluminum surface at 10–11 km/s); for impacts outside this regime (such as impacts of steel or tantalum objects) the variations can range up to a factor of about 15 or more. Since the main basis of models for the sizes of small debris particles comes from these data, improving scaling laws may be an effective means of improving models of the debris environment. Recently there has been some effort toward this end. LDEF debris experiments emphasized the need to acquire chemical data on impact features; such data helped to improve damage models as well as estimates of the debris population. LDEF experiments also resulted in the development of a consistent set of physics-based scaling laws for all velocities and ductile materials (Watts et al., 1993). Reevaluation of historical data using these scaling laws could result in a more reliable data set on which to base environment models for small debris. These physics-based scaling laws, however, require a thorough understanding of both the materials (impactor and target) and the impactor shape involved in the impact, so they may not be applicable to many experiments.

Active Measurements

Active detectors have been used to detect meteoroid impacts since the early days of space activity. For many years, Salyut space stations had active impact detectors (Kuzin, 1993), as did Explorers 16, 23, and 46, and the Pegasus series of spacecraft in the 1960s (Mulholland, 1993). There are a wide range of active detectors, from simple impact detectors to complex chemical composition sensors. The simplest and cheapest detectors (and the ones most able to be made into large area detection systems) are acoustic, piezoelectric, pressurized cell, and capacitive discharge impact detectors. These and other simple impact detectors emit a signal when impacted or perforated. There are also many complex detectors (such as plasma detectors, plasma charge separation systems, optical photometers, and chemical and spectrum analyzers) that return a wide range of data regarding the impactor (Atkinson et al., 1993).

Active detectors are able to acquire characterization data that cannot be obtained by passive means. For example, time-dependent measurements of the environment can be made only with active detectors. Such measurements provide the necessary data for monitoring short-term changes in the environment as well as for determining and modeling the dynamics of environmental processes such as formation, distribution, target interactions, and orbital decay. The capabilities of active detectors were made clear by LDEF's Interplanetary Dust Experiment, which used very simple active impact detectors (semiconductor capacitors that discharge on impact) to make time-specific measurements of the debris environment that led to the first detection and monitoring of concentrated clouds of small debris particles (Mulholland et al., 1991). Active sensors would also be required for potential future missions such as the detection of collisions through measurements of the flux of small debris (Potter, 1993).

A wide range of different types of active detectors can be deployed together to maximize the data gathered about each debris impact. Such data include information about the number of impacts per unit time and area; the time of each impact; and the velocity, size, and material composition of the impacting particle. Since on-board collection and transmission of data is possible with active detectors, the return of active detectors to the ground is not necessary; this enables the deployment of active detectors at any altitude. If a return to Earth is planned, however, active detectors can be combined with passive detector techniques.

Active detectors typically cost much more than passive detectors. Complex detection systems incorporating multiple active detector techniques to determine impact velocity and impactor composition, such as those flown on the recent Japanese Hiten and German Brem-Sat spacecraft (Hüdepohl et al., 1992) and those planned for the Cassini spacecraft (Ratcliff et al., 1992), can cost hundreds of thousands to millions of dollars to develop and build. However, for specific missions such as detection and monitoring of orbital debris swarms, simple and relatively inexpensive active detector systems can be built and deployed. For example, the Clementine 1 interstage adaptor incorporated capacitive discharge impact detectors similar to those flown on LDEF. Approximately 0.14 square meter of exposed detector area was placed around the circumference of the adapter, which was discarded in a highly elliptical orbit around the Earth. This Orbital Meteoroid and Debris Counter (Kinard, 1993) experiment, which had a total weight of approximately 0.5 kg, operated until the interstage adapter reentered the atmosphere in May 1994. The total design, development, and integration cost of this experiment was $200,000.

Active detectors can have a variety of other limitations, depending on the type of detector. First, complex active detectors are often inherently limited to a few tens of square centimeters of exposed area, can have high masses (in the tens of kilograms), and require large volumes to contain the instruments and associated electronics. In addition, active detectors can suffer from problems with data interpretation and can require many calibration tests. The majority of recent developmental work on active detectors has focused on reducing cost and weight for a given level of performance (e.g., Mulholland, 1993), the development of combined detector systems, and better calibration of currently available detectors (e.g., Kassel and Wortman, 1994).

Extending the Range of In Situ Detectors

In situ detectors have the potential to be used to better characterize the population of medium-sized debris particles. As discussed previously, debris particles a few millimeters in diameter (the lower end of the medium-sized debris range) are very difficult to characterize with even improved ground-based sensors, and remote sensing of such particles in LEO would be a difficult and probably costly effort. The basic problem with detecting these objects via in situ techniques is that (as is discussed in Chapter 3) the flux of medium-sized objects is much lower than the flux of small objects. Medium-sized debris will thus impact a given sensor much less often than will small debris, producing much less data to analyze.

Either very large or very long duration in situ sensors, however, have the potential to provide an effective means of sampling the medium-sized debris population by exposing a large enough surface area over a long enough time for it to be impacted by the relatively sparse flux of these particles. There are difficulties with very long duration missions, however: they would obviously not provide data for some time, and their data would be less valuable because they would represent the average flux over a long period of time. Very large detectors may thus be the best means to conduct in situ impact sampling of the population of medium-sized debris up to a few millimeters in diameter (while also providing a great deal of data on the small debris population).

Historically, the available exposure area has been limited such that the largest particle diameters detected by in situ detectors to date have been approximately 1 mm (on LDEF; See et al., 1990). However, proposals have been put forward for achieving much larger exposure areas (Kuzin, 1993; Strong and Tuzzolino, 1989). Both of these proposals depend on the use of thin-film active detectors that generate a signal when perforated. The proposal by Strong and Tuzzolino recommended development of a spacecraft with tens to hundreds of square meters of detector area in multiple large deployable arrays (similar to deployable solar power panels). Based on current predictions of the medium-sized and small debris flux, these large detectors should collide with a few particles as large as 1 mm in diameter (as well as numerous particles smaller than 1 mm) annually in LEO (Kuzin, 1993).

The Kuzin proposal recommended the in-orbit modification of the Progress-M cargo spacecraft used by Russia to support its Mir space station operations. The Progress-M would be modified by space station cosmonauts, drawing on Russian experience in extravehicular activities and the construction of large deployable structures. The modifications would provide power modules and a large deployable detector array (similar in design to deployable communication and radar antennas) that could extend several thousands or tens of thousands of square meters of detector area at space station altitudes. A few debris particles 5 to 10 mm in diameter, as well as numerous smaller particles, are expected to collide with (and thus be detected by) such detector areas annually (Kuzin, 1993).

These large-detector concepts are intriguing and technically achievable today, but may be less cost-effective or feasible than other types of space-based sensors previously discussed and yet not provide more meaningful data. The primary disadvantage of these large active array proposals is the cost of the detectors. The feasibility of ensuring a reasonable orbital lifetime for the detectors at low altitudes (given their potentially high area-to-mass ratios) also needs to be studied. Finally, the potential hazard to other spacecraft from these large arrays may be a problem.

STRATEGIES TO MEASURE THE DEBRIS ENVIRONMENT

Figure 2-3 depicts measurements of the LEO debris environment made since 1980. The major gaps that exist in the altitude and size range data are apparent, as is the intermittent nature of most of the data gather-

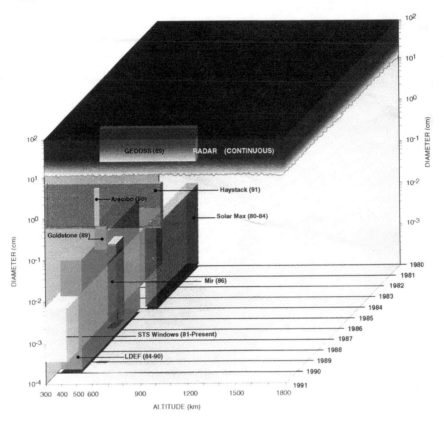

FIGURE 2-3 Orbital debris characterization data—diameter versus altitude versus year. SOURCE: Kaman Sciences Corporation.

ing. (There are actually additional gaps in the data that do not appear in the figure, such as the paucity of data on debris—particularly small and medium-sized debris—in low-inclination orbits.) The haphazard nature of the data is a result of the fact that most measurements of the debris environment to date were not part of an overall strategy to understand the environment but rather were gathered whenever measurement opportunities arose.

Further ad hoc experiments to measure the debris environment will add to our knowledge of debris, but cost-effective characterization of the debris environment (including understanding the time- and altitude-variant nature of the debris population, the sources of small debris, and the collision hazard in widely used orbital regions) will require experiments designed specifically to address these questions. However, there is currently no national or international strategy for implementing experiments to maximize our knowledge of the debris environment. Such a strategy would prioritize the altitude, size, and inclination regimes of highest interest for data collection and would identify the data (such as composition and size, or orbital eccentricity) that are of interest within each regime. This strategy could provide guidance about which detection systems would be most worthwhile to deploy on a given spacecraft and which ground-based sensors could be developed or tasked to observe particular debris size and altitude ranges.

MODELING ORBITAL DEBRIS

Models of the orbital debris population are needed to fill in gaps in the existing measurement data, to interpret new data, and to project the characteristics of the future debris environment. There are two major classes of debris models in use today. Population characterization models take information about the orbital elements and other characteristics of space objects and convert them into measurable parameters such as flux, detection rate for an instrument, or collision velocity. More complex models are used to understand the future growth in the debris population. These model types are not entirely distinct; the output of a model of one type is often used as the input for a model of the other type.

Population Characterization Models

Population characterization models convert data on the orbital elements and other characteristics of space objects into measurable parameters, such as flux, detection rate for an instrument, or collision probability. This conversion is necessary both to help researchers interpret data collected in experiments that sample the uncataloged orbital debris environment and to aid designers in determining the debris hazard to their spacecraft.

Different types of population characterization models have different degrees of uncertainty. Determining the probability that an object in a certain orbit will pass through a particular area of space, for example, requires few assumptions (Kessler, 1981a). Consequently, the average rate at which a given set of objects in known or assumed orbits will pass through a sensor's field of view or impact another object can be calculated with an accuracy of a few percent. However, attempts to determine other characteristics, such as size or albedo, of objects detected by a sensor will usually have a greater uncertainty, due to the variables that contribute to the sensor's signal return.

The application of additional population characterization models, however, can reduce these uncertainties. For example, the diameter of an impact crater on a spacecraft surface is related not only to the impacting object's mass but also to its velocity (speed and direction), which is related to the object's orbital characteristics. Population characterization models can thus be employed to predict a probability distribution of velocities from an assumed orbit distribution, which can then be used to create a probability distribution of particle masses. The impacting object's mass can then be estimated from this probability distribution. A similar method can be used to estimate the size of space objects detected by a telescope. For a telescope, the brightness of an observed object is a function of the object's size, optical properties, and orientation as it passes through the telescope's field of view. In this case, population characterization models can use expected distributions of optical properties and orientations to convert the measured distributions of brightnesses into a distribution of probable sizes.

NASA's "Engineering Orbital Debris Model" (Kessler et al., 1991), and the ESA engineering model (Sdunnus and Klinkrad, 1993) are examples of a particular type of population characterization model used by spacecraft design engineers. These models predict the flux of orbital debris that might strike a spacecraft during its lifetime as a function of debris size and velocity for various spacecraft orbital altitudes and inclinations. Although such models are based primarily on measurements of the orbital debris environment, they use the results of more complex models to extrapolate these measurements.

This type of model also serves as a "reference model" and is used to compare measurements and evaluate relative hazards. There are currently no recognized standard population characterization reference models; researchers and designers must rely on models that have not undergone peer review or that may not contain the latest data. This can potentially lead experimenters to interpret their data improperly or spacecraft designers to improperly assess the hazard to their spacecraft.

Models of the Future Debris Population

The earliest models used to predict the future orbital debris environment (Kessler and Cour-Palais, 1978; Kessler, 1981b; Su and Kessler, 1985) built on the population characterization models and combined breakup models with atmospheric drag models to predict the environment in the 1980s and beyond. These relatively simple models predicted an environment in the 1990s that is not greatly different from that being measured today. Currently, more complex models are used to predict the growth in the orbital debris population. Such models combine a traffic model, a breakup model, and an orbit propagation model to predict possible future orbital debris population states. Two such models in common use today are NASA's EVOLVE (Reynolds, 1993) and the University of Braunschweig's CHAIN (Rex and Eichler, 1993). These models take estimates of the current space object population, add new debris from various sources (e.g., collisions, explosions, mission-related debris), and propagate the orbits of these objects over time to create a static description of the debris population at a selected time in the future. (The predictions these models make about the future debris environment are discussed at length in Chapter 8.)

Each of the component models that goes into such models as EVOLVE and CHAIN has its own characteristics and uncertainties. A *traffic model* keeps track of spacecraft, rocket bodies, and any associated debris launched into orbit by recording when these objects are placed in orbit, their sizes and masses, and their initial orbital elements. Some of these objects will break up into smaller fragments or degrade and release smaller debris. A *breakup model* describes the number of fragments generated in a breakup, as well as the changes in velocity that place them into slightly different orbits. An *orbit propagation model* then determines how the orbits of both intact space objects and space object fragments change as a function of time.

Traffic Modeling

The growth and evolution of the Earth-orbiting space object population will be influenced in large measure by the frequency and character of future space operations. Space traffic models, coupled with propagation and breakup models, predict the magnitude and nature of these operations and their effect on the LEO and HEO space object populations. Traffic models must account for (1) all objects (e.g., spacecraft, rocket bodies, mission-related debris) to be placed into Earth orbit; (2) the apogee, perigee, and inclination of each object's orbit; (3) the size and mass of each object, (4) any planned reorbiting or deorbiting maneuvers at the end of an object's functional lifetime; and (5) any stored energy left in the rocket body or spacecraft that may cause it to explode.

Ideally, space traffic models should look far enough into the future to assess the impact of actions to curb the growth of the total space object population. Predicting even the overall level of space activities over such a time frame, however, is often futile, since very few national or commercial space programs have credible long-range plans extending for more than 8 to 10 years, and even these plans are affected by programmatic, technical, and economic trends; changing national and market requirements; and advances in technology. As a further complication of the problem, it is important to know the population in each orbital region, so that low traffic estimates in one altitude region of the model do not offset unanticipated missions in another altitude region. For all of these reasons, space traffic models have historically been poor predictors of future activities. Nevertheless, scenarios of potential levels of future activity can be developed and used to evaluate the influence of future launch activity on population.

Breakup Models

Breakup models are used to characterize the fragments generated in space object breakups. The results of these models are typically used to estimate existing debris populations and to predict the future population. Most breakup models use the type and amount of energy causing the breakup of a space object of a given mass to estimate the resulting fragment distribution. The most useful breakup models are semi-empirical and incorporate the laws of physics as well as existing data on breakups in their calculations. However, there are two major difficulties involved in developing an accurate breakup model. First, no "typical" amount of debris is generated in an explosion or collision, since there are many different causes of explosions and many different types of collisions (e.g., two spacecraft colliding head-on will produce more debris than a collision between a 10-cm fragment and a spacecraft's solar array). Second, and perhaps more problematically, there are very few data on which to base breakup models.

Few experiments have been conducted to improve breakup models; most available data have been obtained as a byproduct of experiments with other objectives. Explosion data have been gathered from such sources as an accidental Atlas missile explosion, munition explosion tests (Bess, 1975), and explosions in orbit, although recently, some ground-based explosion tests have been conducted specifically to determine the velocity and mass distributions of explosion fragments (Fucke, 1993). Data on collisions are also limited; for many years, the primary sources of such data were the pioneering work of Bess at the NASA Langley Research Center in 1975 and several series of tests performed for the U.S. military during the late 1970s and early 1980s. Debris from the military tests were examined for NASA in the 1980s explicitly to refine the foundation of satellite impact breakup models. The deliberate on-orbit collisions of P-78 and D-180 in the mid 1980s added to this database, though no significant data are available on the smaller (untrackable) fragments produced in these tests.

Recently, however, more complete data on the fragments created in a collision-induced breakup were acquired from tests specifically designed to improve breakup models. In these tests, the U.S. Defense Nuclear Agency shot a 150-gram projectile at 6 km/s into parts of an actual space-

---

BOX 2-6 Modeling Debris Clouds

One specialized type of breakup model focuses on the dynamics of the debris clouds formed following a collision or explosion in orbit (Chobotov, 1990). Although these models do not contribute significantly either to estimates of the current population or to the understanding of the long-term debris population, they can be useful in predicting the short-term hazard to spacecraft in orbits near where a breakup occurred. Such information is particularly useful for designers of spacecraft constellations, who are interested in ensuring that the breakup of one spacecraft will not overly endanger other spacecraft in the constellation.

---

craft and into a full-scale spacecraft model (Hogg et al., 1993). Unfortunately, analysis of the data from these tests was not completed due to a lack of funding. Consequently they have not resulted in any significant improvements in most breakup models, although the tests did demonstrate that breakup models that predicted few small fragments were incorrect. NASA has recently contracted with Kaman Sciences Corporation to complete the analysis of these tests.

These data, particularly the data from the in-orbit breakups, shed light mostly on the characteristics of the larger debris produced. Only the largest fragments of a breakup in orbit can be tracked, although fairly accurate velocity and area-to-mass ratios can be determined for these fragments. Even in ground tests, often only the larger fragments are recovered, since a great amount of work is required to recover the smaller pieces. As a result, the amount and the velocities of smaller debris produced in breakups are not well known.

Propagation Models

Orbit propagation models predict how the orbits of space objects change as a function of time. This information is used for two major purposes: determining the location of particular space objects in the relatively near term (typically over a period of a few days or less for purposes of collision avoidance or reentry predictions) and making long-term (typically over a period of years) predictions about the debris environment. The short- and long-term propagation tasks have some common characteristics, but each also faces unique challenges.

Both short- and long-term propagation models must take into account the various forces acting on space objects in Earth orbit. As described in Chapter 1, these include atmospheric drag, solar radiation pressure, gravitational perturbations by the Sun and Moon, and irregularities in the gravitational field of the Earth. Fortunately, few objects in Earth orbit are affected significantly by many of these forces; the particular forces relevant to each object depend on the object's orbit and area-to-mass ratio. Since accurate orbit propagation models that include all forces acting on an orbiting object can be very computation intensive, most models take into account only the forces that most strongly affect the space objects in a particular orbital region. (For example, in LEO, where orbital inclination does not change significantly with time, the long-term propagation task is reduced to determining the changes in orbital perigee and apogee due to atmospheric drag.)

Accurate short-term deterministic propagation models require that the forces on an object be known and predictable. The inherent unpredictability in atmospheric drag (discussed in Chapter 1) thus introduces error into the predictions of short-term deterministic propagation models for objects in low LEO orbits (less than about 500 km). Accurate deterministic predictions in this region for tasks such as collision warning, which require a high degree of accuracy and propagation of at least a significant fraction of a day, can be achieved only by making repeated observations with increasing calculation fidelity as the time to impact decreases. The Russian SSS uses such an approach to solve actual tasks in debris-related contingencies (e.g., space objects about to reenter). Its approach employs short-term density prediction models utilizing (in addition to knowledge of solar and geomagnetic activity) data on the current drag experienced by other space objects to specify atmospheric density.

Uncertainty in the day-to-day atmospheric drag is not such a problem for long-term propagation modeling in LEO, both because much of the uncertainty can be averaged over time and because long-term models are not as concerned with objects in the orbits most affected by atmospheric drag (which tend to reenter the atmosphere fairly rapidly). The long-term uncertainty in atmospheric drag, however, still limits the fidelity of long-term propagation models in LEO. If solar and geomagnetic activity are known, long-term atmospheric density models are accurate to within about 20%. However, atmospheric density can vary by more than a factor of 10 over the 11-year solar cycle, and the level of future solar cycles is unpredictable. Consequently, only very simple LEO propagation models are normally justified for long-term space object population studies.

Although atmospheric drag ceases to be a factor above LEO, space objects at higher altitudes are influenced by solar radiation pressure, lunar and solar perturbations, and irregularities in the Earth's gravity. These can affect an orbit's inclination and eccentricity as well as its apogee and perigee altitude, so more complex propagation models are required to obtain predictive accuracy. Although such models exist and are capable of providing sufficient accuracy for long-term modeling, they require a very large amount of computation. New hardware, however, is making the calculation-intensive computations much more feasible. It is not yet clear what approximations could be made to enable the creation of accurate long-term HEO propagation models that do not require a large computational capability.

Short-term propagation modeling (for purposes of collision avoidance, etc.) at high altitudes is difficult because of the problems inherent in tracking objects at such distances. One problem is that only very large objects at those distances from the Earth can be detected by current space surveillance sensors. Another is the fairly large uncertainty in the exact position of detected objects. Although short-term predictions have been made for GEO since the 1970s, and avoidance maneuvers have even been carried out based on this information, the uncertainty in the exact position of GEO objects means that the number of false alarms was probably high.

FINDINGS

Finding 1: The U.S. and Russian space surveillance networks are able to detect objects down to a size of about 10 cm in LEO. Increasing fractions of larger objects are tracked so that the LEO debris environment in the size range greater than 20 cm is adequately characterized by the catalogs. However, both catalogs underrepresent objects in highly elliptical orbits, low-inclination orbits, and high-altitude circular orbits. As the orbital altitude increases, the minimum size of objects cataloged grows, until at GEO not even all objects with a diameter greater than 1 meter are tracked.

Finding 2: A number of approaches could be used to improve on current space object catalogs. Sharing catalog data between nations would improve our understanding of the magnitude and distribution of the population of large space objects. A network of new short-wavelength radars would be required to catalog LEO debris significantly smaller than that currently being tracked. Catalogs of large objects in regions above LEO, where data are particularly sparse, could be improved with increased use of large-aperture or CCD-equipped optical sensors. Further analysis is needed to determine whether sharing data from national space object catalogs would result in an improved combined catalog.

Finding 3: In situ direct sampling techniques can detect particle sizes up to about 1 mm in LEO, but the population of medium-sized debris is sufficiently sparse that very large collection areas would be required to obtain a statistically meaningful sample. Ground-based remote sampling has been, and will remain for some time, the most effective means of measuring debris in the medium size ranges.

Finding 4: There has been no systematic approach to sampling space for orbital debris; most sampling to date has been performed when the opportunity arose, resulting in a series of investigations that studied a limited region of space over a limited amount of time. There is a need for national or international strategies to help prioritize detector development, deployment, data collection, and analysis of historical and new data. Such strategies are necessary to gain a better understanding of the sources of small and medium debris and the variations in these populations with respect to altitude, inclination, and time.

Finding 5: Population characterization models can be used by spacecraft designers to estimate the debris hazard to their spacecraft. Debris researchers can use them to integrate available data and to provide a framework for predicting the results of future measurements. As new data become available, existing models should be revised to produce a comprehensive, standard, peer-reviewed reference model.

Finding 6: Models predicting the future space object population in Earth orbit draw on traffic, breakup, and orbit propagation models. These component models have large inherent uncertainties; as a result, many characteristics of the future debris population cannot be predicted with precision. Experience to date with such models has, however, been fairly positive; relatively simple models from the late 1970s and early 1980s predicted an environment in the 1990s that is not greatly different from that being measured today.

REFERENCES

Atkinson, D.R., J.D. Mulholland, A.J. Watts, S.L. Lapin, and J.D. Wagner. 1993. Meteoroid and Debris Monitoring: An Industry Summary. Contract Final Report. Contract Number 959626. Pasadena, California: NASA Jet Propulsion Laboratory.

Batyr, G., S. Veniaminov, V. Dicky, V. Yurasov, A. Menshicov, and Z. Khutorovsky. 1993. The current state of the Russian Space Surveillance System and its capability in surveying space debris. Pp. 43–47 in Proceedings of the First European Conference on Space Debris, Darmstadt, Germany, 5–7 April 1993. Darmstadt: European Space Operations Center.

Batyr, G., S. Veniaminov, V. Dicky, S. Kravchenco, and V. Yurasov. 1994. Some Preliminary Results of ODERACS Experiment. Paper presented at U.S./Russia Orbit Determination and Prediction Workshop, Washington, D.C.

Bendisch, J., J.P. Hoffmann, R. Liebscher, and F. Rollenhagen. 1993. Detection of space debris by the use of space-based optical sensors. Pp. 91–97 in Proceedings of the First European Conference on Space Debris, Darmstadt, Germany, 5–7 April 1993. Darmstadt: European Space Operations Center.

Bess, T. D. 1975. Mass Distribution of Orbiting Man-Made Space Debris. NASA TN D-8108. Washington, D.C.: National Aeronautics and Space Administration.

Chobotov, V. 1990. Dynamics of orbital debris clouds and the resulting collision hazard to spacecraft. Journal of the British Interplanetary Society 43:187–195.

Clark, J., and E. Pearce. 1993. Briefing presented for the National Aeronautics and Space Administration/Department of Defense Orbital Debris Workshop, Colorado Springs, Colorado, September 1–3.

de Jonge, A.R.W., 1993. IR detection of space debris. Briefing presented to the National Research Council Committee on Space Debris Workshop, Irvine, California, November 18.

Fucke, W. 1993. Fragmentation experiments for the evaluation of the small size debris population. Pp. 275–280 in Proceedings of the First European Conference on Space Debris, Darmstadt, Germany, 5–7 April 1993. Darmstadt: European Space Operations Center.

Goldstein, R., and L. Randolph. 1990. Rings of Earth Detected by Orbital Debris Radar. JPL Progress Report 42–101, May 15. Pasadena, California: NASA Jet Propulsion Laboratory.

Henize, H., and J. Stanley. 1990. Optical observations of orbital debris. AIAA-90-1340. AIAA/NASA/DoD Orbital Debris Conference, Baltimore, Maryland, April 16–19. Washington D.C.: American Institute of Aeronautics and Astronautics.

Hogg, D.M., T.M. Cunningham, and W.M. Isbell. 1993. Final Report on the SOCIT Series of Hypervelocity Impact Tests. Wright Laboratory-TR-93-7025. Dayton, Ohio: Wright Laboratory.

Hüdepohl, A., H. Iglseder, and E. Grün. 1992. Analysis of the Results of Two-Year Operations of the Munich Dust Counter—A Cosmic Dust Experiment on Board the Satellite HITEN. International Astronautical Federation paper 92-0561. Paris: International Astronautical Federation.

Kassel, P.C., and J.J. Wortman. 1994. Metal-oxide-silicon capacitor detectors for measuring micrometeoroid and space debris flux. Paper submitted to the AIAA Journal of Spacecraft and Rockets.

Kessler, D.J. 1981a. Derivation of the collision probability between orbiting objects: The lifetime of Jupiter's outer moons. Icarus 48:39–48.

Kessler, D.J. 1981b. Sources of orbital debris and the projected environment for future spacecraft. Journal of Spacecraft and Rockets 18(4):357–360.

Kessler, D.J. 1993. Orbital debris environment. Pp. 251–262 in Proceedings of the First European Conference on Space Debris, Darmstadt, Germany, 5–7 April 1993. Darmstadt: European Space Operations Center.

Kessler, D.J., and B.G. Cour-Palais. 1978. Collision frequency of artificial satellite: The creation of a debris belt. Journal of Geophysical Research 83(A6):2637–2646.

Kessler, D.J., R.C. Reynolds, and P.D. Anz-Meador. 1991. Space Station Program Natural Environment Definition for Design. NASA SSP 30425, Revision A. Houston, Texas: NASA Johnson Space Center.

Kinard, W.H. 1993. Description of the orbital meteoroid and debris counting experiment to fly on the CLEMENTINE inter-stage adapter spacecraft. P. 148 in NASA Conference Publication 10120, Third LDEF Post-Retrieval Symposium Abstracts. Hampton, Virginia: NASA Langley Research Center.

Kuzin, G.A. 1993. Meteorite and Man-Made Microparticle Impact Detection Methodology and Equipment on the Space Stations "SALYUT" and "MIR." Briefing presented to the National Research Council Committee on Space Debris Workshop, Irvine, California, November 18.

McDonnell, J.A.M., and K. Sullivan. 1992. Hypervelocity impacts on space detectors: Decoding the projectile parameters. Hypervelocity Impacts in Space, 39. Canterbury, United Kingdom: University of Kent at Canterbury Unit for Space Sciences.

Mehrholz, D. 1993. Potentials and limits of space object observations and data analysis using radar techniques. Pp. 59–64 in Proceedings of the First European Conference on Space Debris, Darmstadt, Germany, 5–7 April 1993. Darmstadt: European Space Operations Center.

Mulholland, J.D. 1993. Synoptic monitoring of orbital debris (SYNMOD): A progress report on current and future applications. P. 149 in Proceedings of the First European Conference on Space Debris, Darmstadt, Germany, 5–7 April 1993. Darmstadt: European Space Operations Center.

Mulholland, J.D., S.F. Singer, J.P. Oliver, J.L. Weinberg, W.J. Cooke, N.L. Montague, J.J. Wortman, P.C. Kassel, and W.H. Kinard. 1991. IDE Spatio-Temporal Impact Fluxes and High Time-Resolution Studies of Multi-Impact Events and Long-Lived Debris Clouds. NASA CR-3134. NASA Conference Publication 3134, LDEF-69 Months in Space: Proceedings of the First Post-Retrieval Symposium, Kissimmee, Florida, June 2–8. A.S. Levine, ed. Hampton, Virginia: NASA Langley Research Center.

NASA (National Aeronautics and Space Administration). 1970. Meteoroid Damage Assessment. NASA Space Vehicle Design Criteria. NASA SP-8042. Washington, D.C.: NASA.

NASA (National Aeronautics and Space Administration). 1990. Report of the Subcommittee on Micrometeor and Debris Protection, Space Station Advisory Council. Washington, D.C.: NASA.

Neste, S., K. Tomiyasu, H. Halsey, R. Grenda, and R. Soberman. 1982. Feasibility Study for Space Debris Detection Concepts Final Report, Contract No. NAS9-16459. Philadelphia, Pennsylvania: General Electric Company Space Systems Division.

Pearce, E.C., M.S. Blythe, D.M. Gibson, and P.J. Trujillo. In press. Space debris measurements: Phase one final report. Proceedings of the 1994 Space Surveillance Workshop.

Portree, D.S., and J.P. Loftus, Jr. 1993. Orbital Debris and New-Earth Environment Management: A Chronology. NASA Reference Publication 1320, December. Linthicum Heights, Maryland: NASA Center for Aerospace Information.

Potter, A.E. 1993. Early detection of collisional cascading. Pp. 281–285 in Proceedings of the First European Conference on Space Debris, Darmstadt, Germany, 5–7 April 1993. Darmstadt: European Space Operations Center.

Ratcliff, P.R., J.A.M. McDonnell, J.G. Firth, and E. Grün. 1992. The cosmic dust analyzer. Journal of the British Interplanetary Society 45(9):375.

Rex, D., and P. Eichler. 1993. The possible long term overcrowding of LEO and the necessity and effectiveness of debris mitigation measures. Pp. 604–615 in Proceedings of the First European Conference on Space Debris, Darmstadt, Germany, 5–7 April 1993. Darmstadt: European Space Operations Center.

Reynolds, R.C. 1993. Orbital debris environment projections for space station. Pp. 337–339 in Proceedings of the First European Conference on Space Debris, Darmstadt, Germany, 5–7 April 1993. Darmstadt: European Space Operations Center.

Sato, T., K. Tanaka, K.I. Ikeda, T. Wakayama, and I. Kimura. 1992. Interpretation of space debris RCS variations observed by the MU radar. Paper presented at the 18th International Symposium on Space Technology and Space Science (18th ISTS), Kagoshima, Japan, May 17–23. Tokyo: ISTS.

Sdunnus, H., and H. Klinkrad. 1993. An Introduction to the ESA Reference Model for Space Debris and Meteoroids. Pp. 343–348 in Proceedings of the First European Conference on Space Debris, Darmstadt, Germany, 5–7 April 1993. Darmstadt: European Space Operations Center.

See, T., M. Allbrooks, D. Atkinson, C. Simon, and M. Zolensky. 1990. Meteoroid and Debris Impact Features Documented on the Long Duration Exposure Facility: A Preliminary Report. NASA JSC #24608. Houston, Texas: NASA Johnson Space Center.

Stansbery, E.G., D.J. Kessler, T.E. Tracy, M.J. Matney, and J.F. Stanley. 1994. Haystack Radar Measurements of the Orbital Debris Environment. JSC-26655 May 20. Houston, Texas: NASA Johnson Space Center.

Strong, I., and A.J. Tuzzolino. 1989. The Space Dust Experimental Spacecraft. Air Force Laboratory Contract Task Report. Contract F29601-87-C-0207, Subtask 03-11. Paper presented to the First Annual Workshop on the Effects of Orbital Debris, Kirtland Air Force Base, New Mexico, March 15–16.

Su, S.Y., and D.J. Kessler. 1985. Contribution of explosion and future collision fragments to the orbital debris environment. Advances in Space Research 5(2):25–34.

Thompson, T.W., R.M. Goldstein, D.B. Campbell, E.G. Stansbery, and A.E. Potter. 1992. Radar detection of centimeter-sized orbital debris: Preliminary Arecibo observations at 12.5-cm wavelength. Geophysical Research Letters 19(3):257.

Utkin, V.F., V.I. Lukyashenko, V.N. Dolgih, V.A. Emelyanov, M.V. Yakovlev, and A.B. Krasnov. 1993. The state of the art in research of technogenic space debris in Russia: Proposals for international cooperation. Pp. 15–18 in Proceedings of the First European Conference on Space Debris, Darmstadt, Germany, 5–7 April 1993. Darmstadt: European Space Operations Center.

Watts, A., D. Atkinson, and S. Rieco. 1993. Dimensional Scaling for Impact Cratering and Perforation. NASA NCR-188259 March 16. Houston, Texas: NASA Johnson Space Center.

3

Debris Population Distribution

As discussed in Chapter 2, a variety of techniques have been developed to characterize the orbital debris environment, but a high level of uncertainty remains in our understanding of the debris population. While extensive data have been acquired on the cataloged population, cataloged objects represent only a small fraction of the debris in orbit; estimates of the populations of uncataloged debris are based on a limited number of sampling measurements tied together with models. Any estimates of the overall debris population are thus uncertain; they are likely to change as new data are acquired. Figure 3-1 presents one estimate of the total number of objects of various sizes in LEO, based on various measurements. Table 3-1 estimates the total orbital debris population in each size range and the fraction of the total mass in orbit contributed by objects in each size range.

LARGE DEBRIS

The best-known segment of the debris population is the population of cataloged large debris. Figure 3-2 is a "snapshot" depiction of the location of all cataloged debris at a particular moment in time. Some features of the distribution of the cataloged debris population can already be seen in this figure, including the concentrations in the GEO ring and in LEO. Figure 3-3 quantifies Figure 3-2 by portraying the approximate spatial density of cataloged objects at various altitudes. Clear concentrations can be seen at less than 2,000-km altitude (LEO), around 20,000 km (semisynchronous orbit), and at 36,000 km (GEO).

63

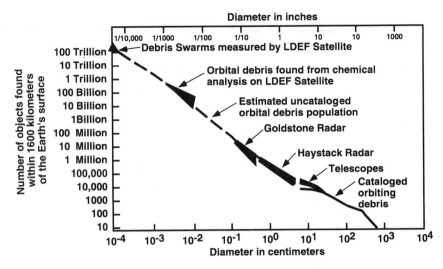

FIGURE 3-1 Number of objects in LEO as estimated from various measurements. SOURCE: National Aeronautics and Space Administration.

These concentrations of higher spatial density are due to large numbers of objects in near-circular orbits at or near these altitudes. The lower background level of spatial density visible in Figure 3-3 at altitudes up to 40,000 km is due to objects in highly elliptical orbits with perigees in LEO and apogees up to 40,000 km. This background spatial density also exists in LEO, where most highly elliptical orbits have their perigee. Most objects in highly elliptical orbits are either rocket bodies that placed spacecraft in semisynchronous orbit or GEO or objects in Molniya-type orbits. Few objects are cataloged in orbits higher than 40,000 km.

Figures 3-4 and 3-5 indicate the distribution of different types of cataloged space objects by mean altitude. At less than 2,000 km, the majority of cataloged objects are fragmentation debris, but at altitudes between 2,000 and 16,000 km, mission-related debris represents the largest frac-

TABLE 3-1 Approximate Orbital Debris Population by Size

Orbital Debris Size Range	Number of Objects	Percentage of Objects > 1 mm	Percentage of Total Mass
Large (>10 cm)	> 10,000	<0.5	>99.95
Medium (1 mm-10 cm)	Perhaps tens of millions	>99.5	<0.05
Small (<1 mm)	Trillions	–	<0.01

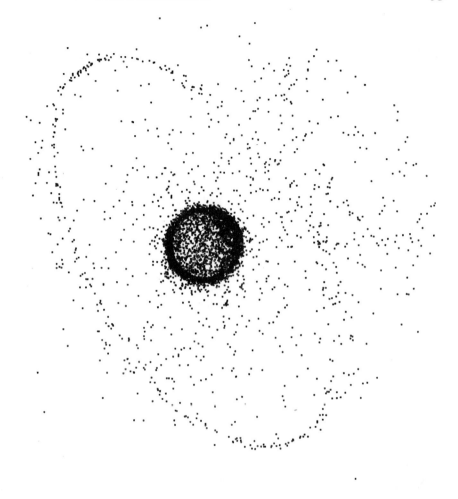

FIGURE 3-2 Cataloged orbital debris. SOURCE: Kaman Sciences Corporation.

tion of cataloged objects; at more than 16,000 km, spacecraft and rocket bodies constitute the majority. This distribution may, however, be due more to the reduced capabilities of Earth-based sensors to detect smaller objects at high altitudes than to any changes in the composition of the debris population. Within the region below 2,000 km, the distribution of cataloged objects by altitude is highly nonuniform, with peaks around 900 to 1,000 km and 1,400 to 1,500 km. Although objects in the lowest-

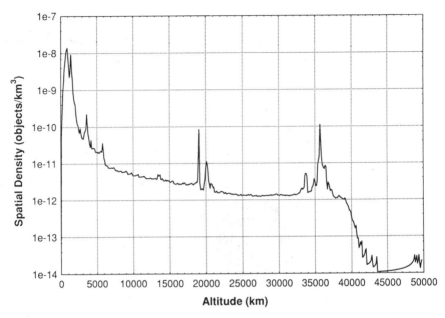

FIGURE 3-3 Spatial density of the 1994 U.S. Space Command Satellite Catalog. SOURCE: Kaman Sciences Corporation.

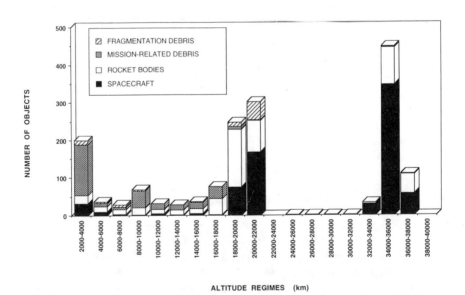

FIGURE 3-4 Low altitude space object population by semi-major axis, 1993. SOURCE: Prepared by Kaman Sciences Corporation based on U.S. Space Command Satellite Catalog.

Erratum: This chart should replace the one appearing in Figure 3-4 on page 66 in *Orbital Debris: A Technical Assessment.*

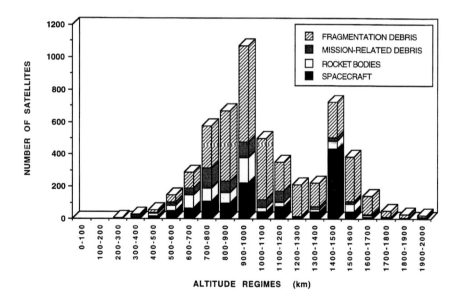

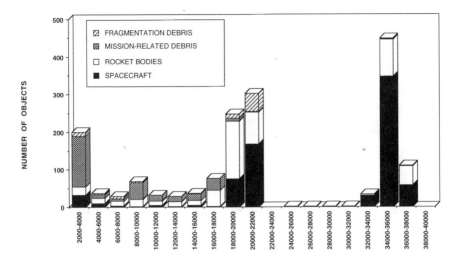

FIGURE 3-5 High altitude space object population by semi-major axis, 1993. SOURCE: Prepared by Kaman Sciences Corporation based on U.S. Space Command Satellite Catalog.

altitude orbits eventually reenter the atmosphere, this population is augmented by objects from decaying higher-altitude orbits.

Except for those in GEO, most cataloged objects are in orbits with fairly high inclinations. This means that relative collision velocities for these objects will be generally higher than orbital velocity. (Collision velocities are discussed in detail in Chapter 4.) Differing orbital inclinations also cause asymmetric distributions in the LEO satellite population by latitude. For example, objects in low-inclination orbits do not contribute to the apparent congestion or bunching of objects in the higher temperate zones, and since few objects are in truly polar orbits (with inclinations of 90 degrees), "holes" in the space object swarm appear over the Earth's poles. (This does not, however, mean that high-inclination orbits will have a lower collision probability; any two circular orbits at the same altitude will intersect at two points, irrespective of their respective inclinations.) Figure 3-6 shows the inclination distribution of cataloged space objects.

Above LEO, spacecraft in orbits at a particular altitude often have similar missions, so both they and the debris associated with them (e.g., rocket bodies, mission-related debris, fragmentation debris) tend to have similar inclinations. These high-altitude, high-inclination orbits include Molniya-type orbits, which typically have inclinations of 63 to 65 degrees

---

BOX 3-1   LEO Communications Constellations

Large constellations of LEO communications spacecraft have been proposed by a number of companies and organizations. These include the Iridium system of 66 spacecraft, the Globalstar constellation of 48 spacecraft, and the Teledesic constellation of 840 spacecraft, among others. Launches of spacecraft for these constellations could begin in the middle to late 1990s. If these constellations are developed, they will add significantly to the population of large objects in LEO.

(though objects in these orbits experience inclination changes of ±5 degrees) and the orbits near semisynchronous altitude, where inclinations are about 55 degrees for U.S. spacecraft and 65 degrees for CIS spacecraft. Space objects in GEO orbits are originally placed in near-zero inclination orbits, but once stationkeeping stops, the inclination of a GEO object's orbit will vary with time.

Most spacecraft in GEO actively maintain inclinations close to zero degrees and remain stationary above a given longitude. However, the orbital planes of nonfunctional spacecraft and other debris, will (due to

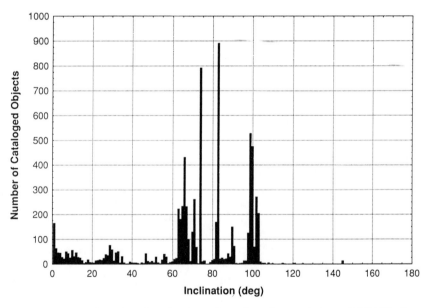

FIGURE 3-6 Inclination distribution of cataloged population. SOURCE: Prepared by Kaman Sciences Corporation, based in part on U.S. Space Command Satellite Catalog, July 1994.

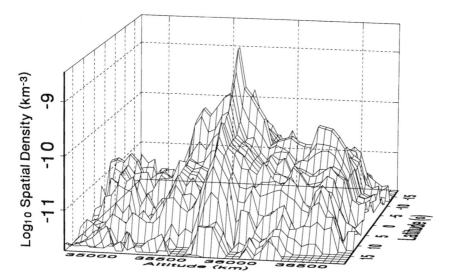

FIGURE 3-7 Geosynchronous spatial density by altitude and latitude. SOURCE: Prepared by Kaman Sciences Corporation based on U.S. Space Command Satellite Catalog, August 1993.

the Earth's oblateness and third-body gravitational perturbations of the Sun and Moon) oscillate around a plane tilted 7.3 degrees from the equator, causing orbital inclination to vary with an amplitude of 14.6 degrees over a period of about 53 years. In addition, the ellipticity of the Earth's equator will cause debris in GEO to drift away from their initial longitudinal position and oscillate around the nearest stable position (either above 75°E or above 105°W) with a period of more than two years. As a result of these forces, the current population of debris in GEO has a mix of inclinations ranging from 0 to 15 degrees (though fragmentation debris from breakups near GEO may have even higher inclinations) and orbital planes that intersect throughout the entire geostationary ring. Figure 3-7 shows the current spatial density of cataloged objects near GEO.

The main distinction between the populations of cataloged and uncataloged large debris is more a product of sensor capabilities than of any inherent differences in the objects. For example, a fragment 30 cm in diameter that would almost certainly be cataloged if it were in LEO would not be cataloged if it were in GEO. However, because spacecraft and rocket bodies in Earth orbit are generally large enough to track, the uncataloged large debris population is composed primarily of mission-related and fragmentation debris. As discussed in Chapter 2, there is known to be a population of uncataloged large debris even in LEO, and the fraction of objects that are not cataloged generally increases with altitude. It is possible that the total uncataloged population of large orbital debris could be as numerous as, or more numerous than, the cataloged population.

MEDIUM-SIZED DEBRIS

The population of medium-sized (approximately 1 mm to 10 cm in diameter) debris is not nearly as well known as the population of large debris. As described in Chapter 2, the only measurements of the medium-sized debris population come from sampling of lower-altitude, higher-inclination LEO orbital regions with ground-based sensors. All other estimates of the size and characteristics of the medium-sized debris population are based entirely on extrapolations.

To a first approximation, it might be expected that medium-sized debris would be found in about the same orbits as large debris, since most medium-sized debris originates from large objects. However, all large objects may not contribute equally to the medium-sized debris population; some types of large object (such as rocket bodies that have been a source of explosive fragmentation) may produce much more debris than others. In addition, as described in Chapter 1, perturbing forces affect different sizes of debris differently. Medium-sized debris, which often has a higher ratio of cross-sectional-area to mass than large debris, will often be more strongly affected by atmospheric drag and thus will experience more rapid orbital decay.

Although there are no measurement data proving the origins of medium-sized debris, most likely the population is composed of fragmentation debris and mission-related objects (since nonfunctional spacecraft and rocket bodies are obviously large debris). The number of medium-sized debris objects detected is large compared to the number of large objects. Since it is generally believed that the majority of this population cannot be mission-related objects, they are most likely fragmentation debris. Consequently, breakup models can be useful tools in estimating some characteristics of the medium-sized debris population. Although there are large uncertainties in predictions of both the number and the initial velocities—and thus orbital parameters—of medium-sized objects ejected in a breakup (as described in Chapter 2), it is known that medium-sized fragments will generally be ejected from a catastrophic breakup with a greater range of initial relative velocities than large breakup fragments; this will place them into orbits with a wider range of altitudes, inclinations, and eccentricities (Johnson, 1985)

Ground-based sensors, particularly the Haystack radar, have provided the most detailed information to date on the population of medium-sized debris objects. Figure 3-8 shows the estimated population distribution of objects detected by the Haystack radar when parked vertically, as compared with the population distribution of objects in the U.S. catalog. Interestingly, the data show that for the region measured, the altitude distribution of medium-sized objects is similar to that of the larger objects included in the U.S. catalog. There are, however, two significant differences: (1) below about 1,000 km the population of medium-sized objects detected by Haystack declines with decreasing altitude faster than the population of large cataloged objects; and (2) around 900 to 1,000 km there is a large peak in the population of medium-sized objects detected by Haystack with no corresponding peak in the population of large cataloged objects. The first difference is consistent with the expectation that medium-sized pieces of debris are more strongly affected by atmospheric drag than larger debris. The peak in the medium-sized population around 900 to 1,000 km, however, points to a source of debris other than previously recorded breakups.

The eccentricity and inclination of many of the medium-sized objects detected by Haystack can also be determined. The data on inclination versus altitude for the objects detected by the Haystack radar are depicted in Figure 3-9. These measurements show that medium-sized de-

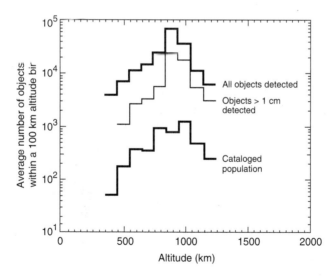

FIGURE 3-8 Estimate of LEO mid-sized orbital debris population from Haystack radar sampling (90 degrees, 547.6 hours), compared to the U.S. Space Command population of cataloged objects. SOURCE: National Aeronautics and Space Administration.

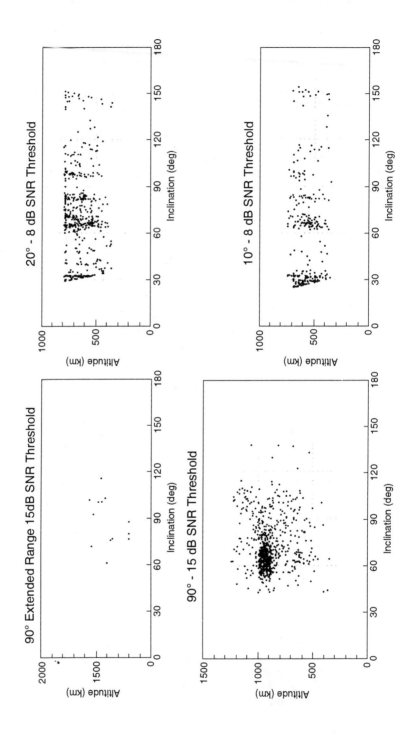

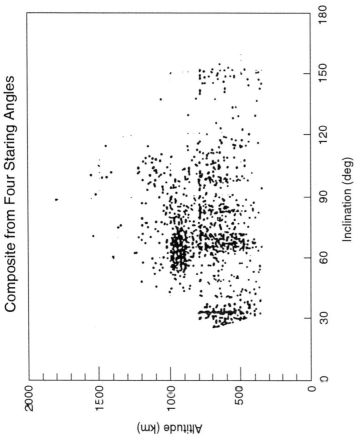

FIGURE 3-9 Altitude *vs.* inclination for detections from various Haystack staring angles. SOURCE: National Aeronautics and Space Administration.

bris is more frequently found in low inclination and eccentric orbits than cataloged large debris and that the large number of objects detected between 900 and 1,000 km are in near-circular orbits with inclinations around 65 degrees (Stansbery et al., 1994). The reported detection of objects with inclinations greater than 110 degrees may be a result of the high uncertainty in determining inclination for objects that are barely detectable (as described in Chapter 2).

As mentioned previously, the Haystack data suggest that there may be major sources of centimeter-sized orbital debris other than previously recorded breakups. The large number of objects in orbits between 900 and 1,000 km with orbital inclinations between 60 and 70 degrees suggests that there is a significant source of debris in this area. If this source were breakups, however, the debris would have been spread over a much wider area than is evidenced by the data. It thus seems possible that some of this debris may be the result of a previously unmodeled source. This possibility is supported by the polarization data from Haystack, which suggests that the objects have relatively smooth and spherical shapes, rather than the irregular shapes that would typically be created in a breakup. A combination of orbital and physical characteristics can be interpreted to suggest that these objects may be tens of thousands of 0.6-2.0 cm diameter liquid droplets of a sodium/potassium coolant leaking from the nonfunctional cores of Russian Radar Ocean Reconnaissance satellites (Stansbery et al., 1995; Kessler et al., 1995). Less evidence exists to suggest the sources of other concentrations of debris not predicted by models (such as the concentration of medium-sized objects detected by Haystack with inclinations between 25 and 30 degrees—another region in which few breakups have been observed [Kessler, 1993]).

SMALL DEBRIS

There is an extremely numerous population of small (<1 mm in diameter) debris particles in Earth orbit. Knowledge of the distribution of these particles comes, as described in Chapter 2, primarily from the examination of returned spacecraft material from such spacecraft as Solar Max and the LDEF and a few active measurements made on the LDEF, the Salyut and Mir space stations, EURECA, and the U.S. Space Shuttle. Since the returned materials and active measurements are all from spacecraft in orbits of 600 km altitude or less, uncertainty remains on how to extrapolate these data to higher altitudes. Some models predict that because of the lessening influence of atmospheric drag, the spatial density of debris smaller than 1 mm should increase with altitude up to at least 1,000 km.

Like medium-sized debris, small debris is all either mission-related objects (e.g. aluminum oxide particles expelled from solid rocket motors) or fragmentation debris (the product of either breakups or surface deterioration). *Aluminum oxide particles from solid rocket motor exhaust* are generally believed to be approximately spherical in shape with a maximum diameter of about 10 microns. These particles are initially ejected from rocket bodies at velocities from about 1.5 to 3.5 km/s, depending on the particle size (smaller particles are generally ejected faster). Most of these particles rapidly reenter the Earth's atmosphere, while others (typically larger particles) are typically sent into a variety of elliptical orbits, depending on where the rocket was fired. *Paint chips and similar products of deterioration* are usually much larger than the aluminum oxide particles, averaging hundreds of microns in diameter. Such debris particles are released from spacecraft or rocket bodies with virtually no initial ejection velocities and thus initially share nearly identical orbits with their parent object. Finally, the *products of breakup* span the entire range of small (as well as medium and large) debris sizes and exhibit a variety of shapes. Small breakup fragments likely experience a larger range of ejection velocities than medium or large fragments, placing them in a wider range of initial orbits.

Perturbing forces affect the orbits of small debris even more strongly than the orbits of medium-sized debris. In particular, the typically larger ratios of cross-sectional-area to mass of small debris means they are more strongly affected by solar radiation pressures and atmospheric drag. Analyses conclude that less than 5% of aluminum oxide particles produced in solid rocket exhaust will remain in orbit after a year (Muller and Kessler, 1985; Akiba et al., 1990), whereas larger particles produced in breakups or from deterioration may remain in orbit for a few years.

Active measurements made during the first year of the LDEF's 1984 to 1990 orbital lifetime first indicated the highly dynamic nature of the small orbital debris environment (though it has since been confirmed by an experiment on the HITEN spacecraft [Münzenmayer et al., 1993]). LDEF's Interplanetary Dust Experiment (Mulholland et al., 1991), which was the only experiment on LDEF that measured the time of impact, showed that most impacts were associated with "orbital debris swarms." That is, the sensors would detect a very large increase in flux (three to five orders of magnitude) lasting for a few minutes. In most cases, these swarms were detected again at nearly the same point in the LDEF orbit. These points slowly changed with time (a characteristic of orbital precession rates), allowing the orbital characteristics of the swarms to be determined. The existence of these swarms suggests that the six-year "average" flux measured by the passive LDEF experiments may in fact be very time dependent, especially for very small debris, of which these swarms mostly consist.

A number of possible sources of these debris swarms have been suggested. One is that the swarms may consist of aluminum oxide particles expelled from solid rocket motors. However, as discussed, such particles experience rapid orbital decay and could not produce swarms lasting for several months, such as those observed by LDEF. It has also been suggested that a spent rocket stage might slowly release sufficient dust to produce the long-lasting swarms (Kessler, 1993). Another possible source might be paint removed by atomic oxygen erosion from objects in highly elliptical orbits. Less than a gram of paint per year removed from a spacecraft would produce a swarm like those detected by LDEF (Kessler, 1990). A final possibility is that the swarms are the result of undetected breakups, perhaps even of a collision. It has been pointed out (Potter, 1993) that the small particles ejected from a hypervelocity impact between a medium-sized debris object and a large object could create a debris cloud having the size distribution of the swarms detected by LDEF.

FINDINGS

Finding 1: The natural meteoroid environment does not pose a serious hazard to most spacecraft in Earth orbit. However, there are orders of magnitude more large orbital debris than large meteoroids in Earth orbit. Although measurements of the medium-sized debris environment are sparse, the population of medium-sized orbital debris also appears to be larger than the population of medium-sized micrometeoroids in the regions of LEO where measurements have been made.

Finding 2: In the limited regions where measurements of the medium-sized debris population have been made, the altitude distribution of the medium-sized objects shows a strong similarity to that of large cataloged objects (except at low altitudes where the influence of atmospheric drag is strong). Measurements of the small debris population, which have been made only at lower altitudes, are so limited that no conclusions about their altitude distribution can yet be drawn.

Finding 3: Because (1) the populations of medium and small debris may change relatively rapidly and (2) our knowledge of these populations comes largely from extrapolations based on a few measurements and models, learning more about the sources of medium and small debris (and improving models with this knowledge) will provide more long-term information about the debris environment than will determining the current spatial density in every orbital region of interest. This is particularly true for the small debris population that (due to short orbital lifetime) may experience drastic changes in a short period of time.

REFERENCES

Akiba, R., N. Ishii, and Y. Inatani. 1990. Behavior of Alumina Particles Exhausted by Solid Rocket Motors. AIAA 90-1367. AIAA/NASA/DOD Orbital Debris Conference, Baltimore, Maryland, April 16–19. Washington, D.C.: American Institute of Aeronautics and Astronautics.

Johnson, N.L. 1985. History and consequences of on-orbit breakups. Advances in Space Research 5:11–19.

Kessler, D.J. 1990. Collision probability at low altitudes resulting from elliptical orbits. Advances in Space Research 10(3-4):393–396.

Kessler, D.J. 1993. Orbital debris environment. Pp. 251–262 in Proceedings of the First European Conference on Space Debris, Darmstadt, Germany, 5–7 April 1993. Darmstadt: European Space Operations Center.

Kessler, D.J., R.C. Reynolds, and P.D. Anz-Meador. 1995. Current Status of Orbital Debris Environment Models. Paper Presented at 33rd Aerospace Sciences Meeting and Exhibit, Reno, Nevada, January 9-12. AIAA 95-0662. Washington, D.C.: American Institute of Aeronautics and Astronautics.

Mulholland, J.D., S.F. Singer, J.P. Oliver, J.L. Weinberg, W.J. Cooke, N.L. Montague, J.J. Wortman, P.C. Kassel, and W.H. Kinard. 1991. IDE spatio-temporal impact fluxes and high time-resolution studies of multi-impact events and long-lived debris clouds. NASA Conference Publication 3134, LDEF–69 Months in Space: Proceedings of the First Post-Retrieval Symposium, Kissimmee, Florida, June 2–8. A.S. Levine, ed. Hampton, Virginia: NASA Langley Research Center.

Muller, A.C., and D.J. Kessler. 1985. The effects of particulates from solid rocket motors fired in space. Advances in Space Research 5(2):77–86.

Münzenmayer, R., H. Iglseder, and H. Svedham. 1993. The Munich dust counter MDC — An experiment for the measurement of micrometeoroids and space debris. Pp. 117–123 in Proceedings of the First European Conference on Space Debris, Darmstadt, Germany, 5–7 April 1993. Darmstadt: European Space Operations Center

Potter, A.E. 1993. Early detection of collisional cascading. Pp. 281–285 in Proceedings of the First European Conference on Space Debris, Darmstadt, Germany, 5–7 April 1993. Darmstadt: European Space Operations Center.

Stansbery, E.G., D.J. Kessler, T.E. Tracy, M.J. Matney, and J.F. Stanley. 1994. Haystack Radar Measurements of the Orbital Debris Environment. JSC-26655. Houston, Texas: NASA Johnson Space Center.

Stansbery, E.G., D.J. Kessler, and M.J. Matney. 1995. Recent Results of Orbital Debris Measurements From the Haystack Radar. Paper Presented at 33rd Aerospace Sciences Meeting and Exhibit, Reno, Nevada, January 9-12. AIAA 95-0662. Washington, D.C.: American Institute of Aeronautics and Astronautics.

U.S. Space Command. 1994. U.S. Space Command Satellite Catalog. Cheyenne Mountain Air Force Base, Colorado: U.S. Space Command.

4

Hazards to Space Operations from Debris

The hazard to space operations from debris is a function of the nature of those operations and the orbital region in which they take place. The orbital region is important because the debris flux encountered by a spacecraft varies greatly with orbital altitude and, to a lesser extent, orbital inclination. The nature of the operations is a factor because the same piece of debris that could cause serious damage to one type of spacecraft might do little harm to a spacecraft with a different configuration or orbital attitude.

The first step in determining the hazard to a spacecraft from orbital debris is to estimate the debris flux for the spacecraft's orbital region. This information can then be combined with information on the spacecraft's configuration and orbital attitude, and with experimental data and models of the damage caused by hypervelocity impacts, to predict the likelihood that debris will cause damage to the spacecraft during its functional lifetime. The accuracy of such a prediction will depend on (1) the degree to which estimates of the debris flux are correct, and (2) the validity of models used to predict impact damage from debris. As discussed in Chapters 2 and 3, the debris flux in any particular orbital region cannot be determined with a great degree of accuracy because of the uncertainty in current assessments of the debris population (particularly the small, medium-sized, and high-altitude populations). As discussed in Chapter 5, the accuracy of damage predictions for debris impacts is also uncertain. Because both of these factors contain uncertainties, any predictions of the debris hazard to spacecraft will also incorporate a significant degree of uncertainty.

79

CHANCE OF DEBRIS IMPACT

The probability that debris will collide with a given spacecraft depends on the spacecraft's size and the debris flux through its orbital region. The effect of spacecraft size on the likelihood of being struck is simple; the chance of impact is directly proportional to the spacecraft's cross-sectional area relative to the debris flux and the amount of time exposed to the environment. The relationship between the probability of collision and the orbital region is far more complex, varying significantly with altitude and to a lesser degree with inclination.

Low Earth Orbit

Although Figure 4-1 oversimplifies the nature of the LEO debris population, it provides a starting point for estimating the debris impact probability for spacecraft in LEO by showing how the flux of debris varies with debris size. The main oversimplification is the grouping of data

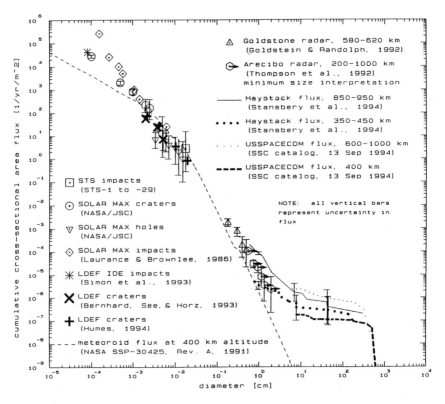

FIGURE 4-1 Approximate debris flux in LEO by object size.

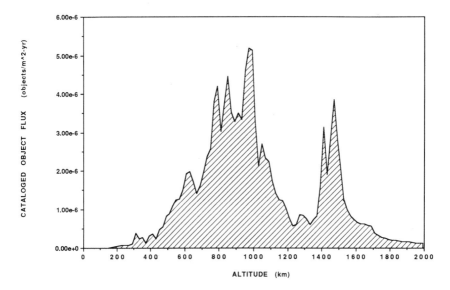

FIGURE 4-2 Flux of LEO cataloged objects, 1994 (assumed velocity 10 km/s). SOURCE: Prepared by Kaman Sciences Corporation based in part on U.S. Space Command data.

acquired at a variety of altitude regimes at different times during a seven-year period. The uncertainty of debris population estimates is, however, reflected by the error bars in the figure.

Figure 4-1 predicts the average number of collisions with different sizes of debris that a spacecraft in a "typical" low Earth orbit will experience in a 10-year orbital lifetime. For example, the probability that a spacecraft in LEO with a cross-sectional area of 10 square meters will collide with an object larger than about 1 cm in diameter over its 10-year functional lifetime can be seen to be somewhere between one in a hundred and one in a thousand. The figure also predicts that the same spacecraft will be struck by about one 1-mm- to 1-cm-diameter particle and somewhere between 100 and 1,000 particles with diameters between 0.1 mm and 1 mm during this time period.

The chances of a spacecraft in LEO being struck by debris can vary significantly from those estimates, depending on the spacecraft's particular orbit. Figure 4-2 shows the variation in flux for cataloged objects in LEO as a function of altitude. While this figure does not show the uncataloged debris flux, the Haystack data have shown that uncataloged debris as small as 0.7 cm in diameter follow a similar distribution to the cataloged flux throughout much of LEO (as discussed in Chapter 3). Figure 4-3 shows an estimate by NASA's EVOLVE model of the flux of large

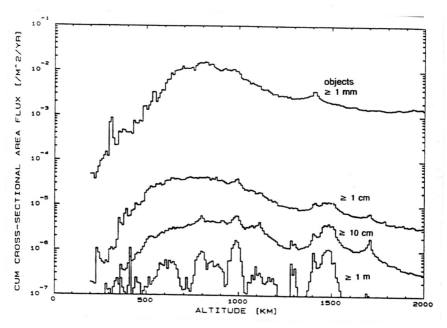

FIGURE 4-3 One model's prediction of the LEO debris hazard. SOURCE: NASA/Reynolds.

and medium objects in LEO. There are some differences between these measurements and predictions, but they all show large variations in the LEO flux (and thus the probability that a spacecraft in LEO will be struck) with altitude. For example, at a typical Space Shuttle orbital altitude of 300 km, the flux of both medium and large debris is about 50 times lower than if it were in an orbit at 1,000-km altitude. (At this altitude, the collision probability will also vary by more than a factor of two due to solar activity.) Since data on the altitude variation of the small debris population do not exist and model predictions vary significantly depending on the sources of these smaller particles, it is unclear whether the chance of being struck by small debris also exhibits such a large variation with altitude.

Collision probability also varies with orbital inclination, although to a much lesser extent than it varies with altitude. The variation with inclination is relatively small because, to a first approximation, two circular orbits at the same altitude will intersect twice per revolution, irrespective of their inclinations. More detailed examinations of the variation of collision probability with inclination (e.g., Kessler, 1981) indicate that collision probability for an orbiting object increases to its maximum value when there are objects in orbits with supplementary inclinations. For example, objects in orbits with inclinations near 80 degrees have a higher collision probability due to the large number of objects in Sun-synchronous orbits with inclinations near 100 degrees. (Similarly, objects in those orbits have higher collision probabilities due to the objects in orbits with inclinations near 80 degrees.) Since there are very few objects in orbits with inclinations greater than 120 degrees, objects with inclinations less than about 60 degrees do not experience a similar rise in collision probability.

Figure 4-4 shows the "average" variation in collision probability with inclination for all altitudes below 1,000 km based on the cataloged population. Because the orbital inclination distribution varies slightly with both time and altitude, this variation in collision probability with inclination will also change as a function of time and altitude. In addition, since measurements made with the Haystack radar suggest that the medium-sized debris population is less concentrated in the higher inclinations than the large debris population (see Chapter 3), the increase in collision

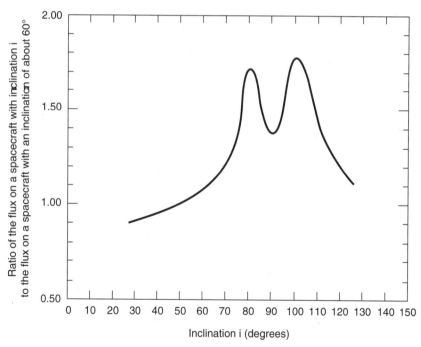

FIGURE 4-4 Average collision probability variation with orbital inclination for cataloged LEO objects. SOURCE: National Aeronautics and Space Administration, based on data from the 1988 U.S. Space Command Catalog.

BOX 4-1   The Meteoroid Environment

Meteoroids revolve about the Sun and steadily rain upon the Earth and on objects in Earth orbit. About 40,000 (±20,000) metric tons of meteoroids enter the Earth's atmosphere each year (Love and Brownlee, 1993). The intensity of this natural environment is often used to establish a threshold of concern for the orbital debris environment.

The onset of space exploration, particularly human space flight, prompted efforts to assess the potential hazard posed to spacecraft by the natural meteoroid environment. Numerous measurements conducted during the 1960s, including the large-area meteoroid detectors deployed by three Pegasus spacecraft in 1965, revealed that the threat of colliding with a meteoroid capable of inflicting significant damage to a spacecraft was remote. (The probability that a square meter of exposed surface in LEO will be struck by a 1-cm-diameter meteoroid during a year in space is about one in a million.) Simple design features are capable of protecting spacecraft against the predominately small and light (average meteoroid density is about 0.5 g/cm^3) particles.

Figure 4-5 shows the estimated meteoroid flux that will be experienced at 500-km altitude. The meteoroid flux varies slowly with altitude due both to shielding by the Earth (which can decrease the flux by as much as a factor of two at low altitudes) and to the Earth's gravity field (which can increase the flux near the Earth by as much as a factor of two) (Kessler, 1972). Just above the Earth's atmosphere, the average velocity of a meteoroid is about 17 km/s; at lunar distances, the average velocity is about 15 km/s. Average meteoroid collision velocities with orbiting spacecraft would be higher by a few kilometers per second, depending on the orbit of the spacecraft (Kessler, 1969).

---

probability with medium-sized objects at high inclinations may not be as great as the estimated increase shown (for cataloged objects) in Figure 4-4.

High Earth Orbits

Estimates of collision probabilities in high Earth orbits are less accurate than LEO collision probability estimates due to the sparse information available on the HEO debris population. (As described in Chapter 2, there are no measurements above LEO of the small debris population, the medium-sized debris population, or even the smaller objects in the large debris population.) It is certain, however, that the chance of collision with cataloged objects is generally much lower in HEO than it is in LEO. As shown in Figure 3-3, the average spatial density of cataloged objects in even the relatively densely populated semisynchronous and geosynchronous orbits is about 100 times lower than the average spatial density of cataloged objects in LEO. In less densely populated high Earth orbits, the spatial density of cataloged objects is often 1,000 times lower

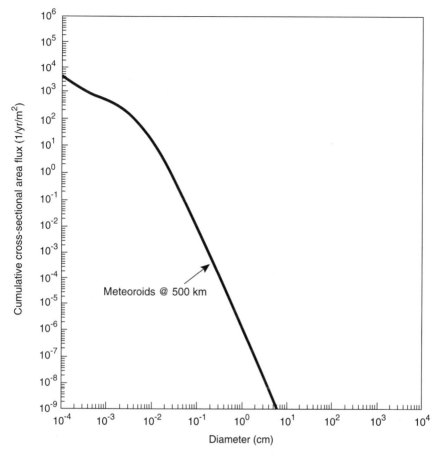

FIGURE 4-5 Meteoroid environment at 500 km altitude. SOURCE: Grün et al., 1985.

than the average LEO spatial density. Although it is unclear how well the distribution of the uncataloged large and medium-sized debris population correlates with the tracked population, it is likely (considering the known sources and perturbing forces) that the average spatial density of these populations is also much lower in HEO than in LEO.

For GEO spacecraft, the chance of collision with cataloged objects decreases sharply with the distance from the geostationary orbit. Figure 4-6 shows how the cataloged space object flux (and thus the probability of collision with a cataloged object) in the GEO region varies as a function of altitude above and below GEO. The flux drops by almost a factor of ten about 50 km above or below the exact geostationary orbit and is approximately two orders of magnitude lower only 500 km above or

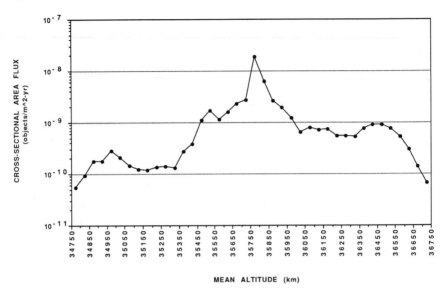

FIGURE 4-6 GEO cross-sectional area flux (0 ± 5 degree latitude, 50-km bins). SOURCE: Kaman Sciences Corporation.

below GEO. Spacecraft in an *inclined* GEO experience about the same flux as shown in Figure 4-6 because spacecraft in such orbits pass through the relatively crowded equatorial geostationary band twice a day, no matter what their inclination.

Because of the difficulty of detecting objects smaller than about a meter in diameter in GEO or even of detecting breakups that could produce smaller objects, the collision hazard from uncataloged debris in GEO is not well known. However, by using the assumption that debris sources in GEO are similar to debris sources in LEO, one model (Kessler, 1993) has predicted that the meteoroid environment will present a greater hazard than the debris environment over the small and medium-sized ranges even if numerous breakups occur. Figure 4-7 illustrates the results of that model.

Space objects in highly elliptical orbits experience different collision probabilities in different parts of their orbit. Objects in Molniya-type orbits experience a very low debris flux through most of their orbit but can spend a small portion of their orbit traveling at high velocities through the relatively intense LEO debris flux. Objects in Molniya orbits will never pass through GEO, and because of the large perigee changes they may experience, many no longer pass through LEO after several years in orbit. However, objects in geostationary transfer orbits will experience the LEO debris flux at the perigee of their orbit as well as the GEO debris flux near their orbit's apogee when the precession of their orbit causes it to pass through GEO. (Figure 4-8 depicts the average time an object in a 27.5 degree inclination GTO will spend in LEO over its orbital lifetime.) Objects in GTO will spend much less time in geostationary orbit than in LEO because orbital precession causes them to pass through the narrow geostationary band only infrequently. This is fortunate because objects in GTO pass through GEO at about 2 km/s, much higher than typical GEO collision velocities.

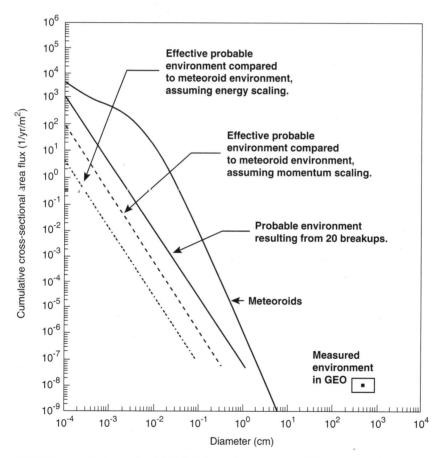

FIGURE 4-7 Estimated orbital debris environment in GEO resulting from satellite breakups. SOURCE: Kessler, 1993.

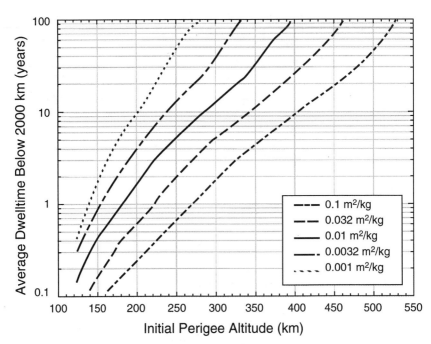

FIGURE 4-8 Average time spent in LEO for GTO, inclination 27.5 degrees. Random choice for initial argument of perigee and right ascension of ascending node of orbit plane. SOURCE: National Aeronautics and Space Administration.

EFFECTS OF DEBRIS IMPACT

Impact Conditions

The damage caused by debris impact depends on the size and velocity of the impacting debris, the configuration and composition of the spacecraft being struck, the component(s) impacted, and the angle at which the impactor strikes the spacecraft. To protect their spacecraft against the debris hazard, designers can calculate typical collision velocities and impact angles and then, if necessary, modify their spacecraft design to protect the areas most likely to be struck by debris. While not perfect, analyses of typical collision velocities and impact angles are based on the known debris population, so they have less uncertainty than many of the other elements factored into debris hazard predictions.

Collision velocities vary with orbital altitude and inclination (see Box 4-2). In LEO, collision velocities vary from almost 0 km/s to greater than 15 km/s. Figure 4-9 shows the calculated proportion of collisions (with cataloged objects) at various velocities as a function of a LEO object's

BOX 4-2   Determining Collision Velocities

Orbital velocities are directly related to altitude—objects in lower-altitude or-bits move faster than objects in higher orbits. Orbital velocity for circular LEO orbits varies from almost 8 km/s (skimming the top of the atmosphere) to about 7 km/s (at 2,000 km). In GEO, orbital velocity is about 3 km/s. The velocity of objects in elliptical orbits varies throughout their orbits. At their perigee, they are moving faster than the local circular orbital velocity, and at apogee, they are moving slower than the local circular orbital velocity.

Impact velocities for objects in circular orbits can vary from nearly 0 km/s for an object striking another object in virtually the same orbit to twice the orbital velocity for a head-on collision. (Collisions with objects in elliptical orbits can occur at even higher velocities.) As the angle at which the two objects' orbits intersect increases toward 180 degrees, so does the collision velocity. If the orbits of the two objects intersect with an angle of greater than 60 degrees, the relative collision velocity will be larger than the objects' orbital velocity.

The impact velocity distribution of the debris flux on a space object is thus influenced by its orbital altitude, eccentricity, and inclination, as well as the eccen-tricity and inclination distribution of objects in intersecting orbits.

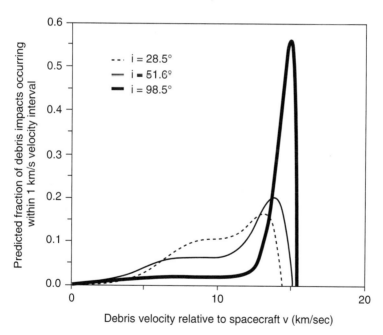

FIGURE 4-9   Calculated collision velocity distribution versus inclination for cat-aloged objects in LEO (averaged over all LEO altitudes).   SOURCE: Calculated from Kessler et al., 1989.

inclination. Clearly, the proportion of high-velocity collisions increases for objects in higher-inclination orbits. If the calculations incorporate the population of objects detected by the Haystack radar in addition to the cataloged population, the plotted variation of collision velocity with altitude looks similar to Figure 4-9, but with slightly lower average collision velocities at all inclinations. In a 51.6-degree-inclination orbit, for example, the predicted average collision velocity with cataloged objects is 10.8 km/s, but the predicted average collision velocity with objects detected by Haystack is 9.2 km/s.

In semisynchronous orbits, orbital velocity is only about 3.9 km/s, so the maximum collision velocity is 7.8 km/s. In practice, however, because most spacecraft in these orbits operate in constellations with inclinations near 60 degrees, the average collision velocity is closer to 4 km/s. In GEO, collision velocities are lower still, both because of the low orbital velocities and because the spacecraft and rocket bodies in GEO are traveling in the same direction and have only minor inclination differences (as discussed in Chapter 3). The long-term average GEO collision velocity due to the various differences in inclination is about 0.5 km/s, much less than the average LEO collision velocity (but still about the speed of a rifle bullet).

The angle at which debris is likely to strike a spacecraft is important for spacecraft designers interested in protecting sensitive components. Figure 4-10 predicts the directions from which debris would impact the

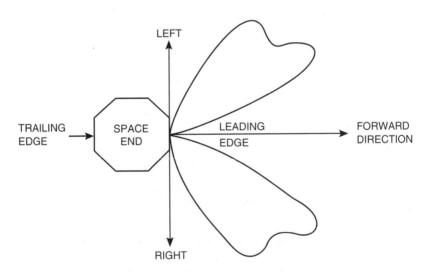

FIGURE 4-10 Direction of orbital debris impact predicted for the LDEF. SOURCE: M&D SIG NASA Model (Chobotov, 1991).

LDEF in its 28.5 degree-inclination LEO orbit, based on the same calculations as Figure 4-9 and with the assumption that the relative velocities in that figure are due to circular orbits. For the same reasons that typical collision velocities change with inclination, the distribution of probable impact angles will be more tightly grouped around the direction of travel for spacecraft in higher-inclination LEO orbits and will be more widely distributed in GEO, where inclination differences between space object orbits are typically small. Debris in highly elliptical orbits may impact the sides and rear of a spacecraft more frequently than debris in circular orbits; such impacts were detected on the rear surfaces of LDEF.

Breakups Due to Debris Impact

Certain high-energy collisions may not just incapacitate a spacecraft, but actually fragment it into many small pieces. Although this distinction may not be important to a spacecraft designer, it is (as discussed in Chapter 8) very important for the future evolution of the debris population. As discussed in Chapter 2, models of such breakups are based on sparse data and contain large uncertainties. Current estimates indicate that complete breakup will occur if the ratio of the impactor's relative kinetic energy to the mass of the object with which it collides is greater than about 40 J/g (McKnight, 1993). For example, a 0.1-kg piece of debris impacting at 10 km/s would probably not completely fragment the Japanese Astro-D spacecraft (420 kg), but a 0.5 kg piece of debris impacting at the same velocity probably would. Of course, the particular geometry of

---

BOX 4-3 Breakup of Space Objects Containing Radioactive Materials

Approximately 55 space objects containing radioactive materials currently orbit the Earth. Although no new spacecraft with nuclear power sources are currently planned for Earth orbit, it is possible that nuclear-powered spacecraft already in orbit may break up as a result of debris impact. Although many such spacecraft were designed so that their nuclear fuel would survive a launch pad explosion relatively intact, all are vulnerable to breakup if involved in a sufficiently high-energy collision.

Radioactive fragments from such a breakup would not interfere with observations by astronomers (only operating reactors produce detectable amounts of x rays or gamma rays), but might reenter the atmosphere sooner than they would have if the spacecraft had remained intact. These small fragments would burn up in the atmosphere, potentially resulting in a slight increase in the background health risk.

an impact is also important, because if the impactor struck a spacecraft's solar panel, for example, it would probably destroy only the panel, rather than the entire spacecraft (although it might cause the spacecraft to start spinning).

Breakup models also predict the number and mass of fragments produced in a catastrophic collision. The mass distribution is also related to the ratio of the impacting object's kinetic energy to the mass of the target space object; as this ratio increases, the number of large fragments produced also increases. For the example above (a 420-kg spacecraft struck by a 0.5-kg piece of debris at 13 km/s), models predict that 50 to 100 fragments with masses of greater than 0.5 kg—massive enough to cause a similar catastrophic breakup—would be produced. Though the total quantity of smaller fragments created in such a collision is more difficult to predict, the number of fragments would increase with decreasing fragment size, totaling millions of medium-sized particles.

These fragments will be ejected at a wide range of velocities, which will place them into a range of new orbits. In general, smaller fragments will be ejected with a wider range of initial velocities than larger ones and thus will be sent into a wider range of new orbits. The velocity of ejected fragments, however, is the most uncertain parameter predicted by breakup models. Figure 4-11 shows an estimate of how the maximum ejection velocities of debris produced in a collision are expected to vary as a function of particle size (Johnson, 1985).

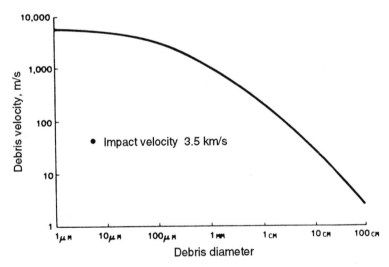

FIGURE 4-11 Maximum ejection velocities of debris as a function of particle diameter. SOURCE: Johnson, 1985.

Structural and Component Damage Caused by the Impact of Debris

In LEO, the impact of medium-sized debris can severely damage or destroy smaller spacecraft or major systems of large spacecraft. Box 4-4 illustrates the destructive force of medium-sized debris traveling at typical LEO collision velocities. In GEO, typical collision velocities are much lower—they are comparable to speeds involved in a midair aircraft collision—so only the largest medium-sized GEO particles are probably capable of causing serious damage.

Hypervelocity impact can cause various modes of damage to spacecraft, including craters, spallations, perforations, and petaled holes and cracks, depending on impact conditions and the configuration of the impacted spacecraft; this damage may result in different failure modes depending on the nature of the spacecraft and the location of the impact. When a piece of medium-sized debris strikes a spacecraft, it can either penetrate the spacecraft's skin or leave a crater on the surface. The impact can cause damage even if it does not penetrate the spacecraft's skin; reflection of the impact's shock wave can cause small particles to spall from the back of the impacted wall. These particles can travel at nearly the velocity of the impacting object, potentially causing serious damage to components inside the spacecraft.

If the impacting debris penetrates the spacecraft's outer skin, its often fragmented or liquefied remnants will travel into the spacecraft and deposit over an area typically significantly larger than the impact hole. The momentum of the impact can cause impulsive damage including buckling and bending of structural components and the transmission of a traveling shock wave through the spacecraft's structure and components. Table 4-1 shows NASA's 1970 assessment of the vulnerability of a spacecraft's subsystems to various modes of hypervelocity impact damage.

The effects of the impact of a 1-cm-diameter aluminum sphere on a

---

BOX 4-4   Energy of High Velocity Objects

The kinetic energy of an object increases with the square of its velocity. The energy of an object moving at 13 km/s (a typical impact velocity in LEO) is roughly equivalent to the energy released by the explosion of 40 times its mass of TNT. For example, a 1-cm-diameter aluminum sphere (which has a mass of about 1.4 grams) moving at 13 km/s has a kinetic energy equivalent to the energy released by the explosion of 56 grams of TNT (about 0.24 MJ); for a 10-cm aluminum sphere, the equivalent is 56 kg of TNT (about 236 MJ).

TABLE 4-1 Potential Failure Modes of Subsystems as a Result of Debris Impact

Probable Critical Types of Failure	Subsystems					
	Pressure Cabins	Tanks	Radiators	Windows	Electronics	Special Surfaces
Catastrophic Rupture	x	x		x		
Detached Spalling	x	x	x		x	
Secondary Fractures			x		x	
Leakage	x	x	x			
Shock Pulse	x			x	x	
Vapor Flash	x					
Deflagration		x				
Deformation			x		x	
Reduced Residual Strength	x	x	x	x		
Fluid Contamination		x	x			
Thermal Insulation Damage	x	x				
Obscuration				x		
Erosion				x		x

x, Subsystem is vulnerable to damage mode

SOURCE: NASA, 1970.

BOX 4-5 Hazards to Crewed Missions

Penetration of the pressure wall of a crewed spacecraft can lead to the loss of cabin pressure, secondary spall impacts on the interior, a light flash, and a pressure pulse. In addition, cracks created by the impact exceeding the critical crack length for a pressurized module can, under some conditions, lead to catastrophic fracture or the uncontrolled mode of crack propagation known as "unzipping."

Astronauts or cosmonauts engaging in extra-vehicular activities are particularly vulnerable to the impact of small debris. On average, debris 1 mm in diameter is capable of perforating current U.S. space suits (Cour-Palais, 1991).

0.5-cm-thick aluminum spacecraft wall at 10 km/s are illustrative of the damage that can be caused by debris impact. Such a collision would fully melt and partially vaporize the impactor and would create a perforation in the spacecraft wall with an outer diameter of approximately 3.3 cm and an inner hole diameter of approximately 2.7 cm. The peak impact stress caused by the expanding liquid projectile and wall material on a component located 2.5 cm behind the perforated wall would be approximately 450 kbar, well above the yield strength for most typical spacecraft structural materials. The peak impact stress decreases with the cube of the distance from the wall, so that the loading on a component 15 cm behind the wall would only be about 2 kbar (still close to the yield strength of commonly used aluminum alloys).

Even small impactors can cause component failures. For example, a particle as small as 0.75-mm diameter impacting 0.5-cm-thick aluminum housing covering a component such as a solar array pointing/steering motor could result in the spallation of the internal housing wall, potentially damaging or jamming the motor. At collision velocities of 10 km/s, particles as small as 1 mm in diameter can perforate a radiator with thin-walled heat pipes (such as those used for space nuclear reactor cooling). If (as is the case with some proposed space nuclear reactor designs) the coolant loop is not designed to allow shutdown of perforated radiator coolant pipes, a loss of coolant could occur.

Surface Degradation Caused by the Impact of Debris

Even if the impacts of smaller debris do not cause structural or component damage, the craters, spallations, and perforations they produce in impacted materials can degrade spacecraft surfaces. (Figure 4-12 shows the surface degradation resulting from an impact into LDEF.) In lami-

FIGURE 4-12 The largest impact crater on LDEF. The central crater measures 5.2 mm in diameter, but ejecta from the crater are spread out over a much larger area. Most experts believe this crater was formed as a result of an impact with orbital debris. SOURCE: National Aeronautics and Space Administration.

nated or multilayered materials, the impact shock can cause delaminations and remove surface coating material far beyond the diameter of the crater. In brittle materials, the impact can initiate cracks extending far beyond the diameter of the crater or perforation. Small debris impacts may also create localized plasmas, which can cause discharges and failures in some components such as electronics or solar arrays. In addition, impact damage may combine with other space environmental effects (such as those caused by atomic oxygen and ultraviolet light) to cause more damage to surfaces than each effect could cause individually.

The spacecraft surface degradation caused by the impact of small orbital debris particles can lead to deterioration of spacecraft performance. With few exceptions, though, performance deterioration models do not exist, and those that do exist are not standardized. One problem is that performance losses are not always directly related to the extent of physical damage caused by debris impact (or to the size of the impacting particles). The effect of surface degradation from debris impact must thus be addressed on a case-by-case basis to evaluate changes in component and system performance.

Optical surfaces are the spacecraft component perhaps most threatened by surface degradation due to debris impact. Impacts by small particles (tens to hundreds of microns in diameter) can significantly increase the light scattered from an optic (Watts et al., 1994). This is particularly important for imaging optics, which usually require very low levels of optical scatter. Small debris impacts into telescope tubes or optical baffles can also degrade optical components by releasing large amounts of particulates (which can temporarily confuse or blind optical sensors) or contaminants (which can affect the scattering of an optical sensor).

Impacts into thermal control components can affect the total available surface area, potentially affecting thermal conduction and radiation and exposing protected areas to the space environment. On LDEF, cratering damage removed approximately 0.26 percent of exposed paint, but the impact-associated front surface spalls increased the total material

---

BOX 4-6   The Effect of Debris on Tethers

Tethers are just beginning to be used in space operations but have a great deal of promise for future applications. Tethers are particularly vulnerable to even small debris because, due to their extremely long and thin shapes, they have large surface areas yet can be severed by a sufficiently large debris impact anywhere along their length.

Based on estimates of the flux of small debris and meteoroids, it has been predicted that a 1-mm-diameter, 20-km-long, single-strand aluminum tether at 500 km will be severed by either meteoroids or debris in an average of three weeks (Kessler, 1984). Either increasing the thickness of the tether or using multiple strands will increase this time. Meteoroid impacts are the primary hazard to tethers thinner than a few millimeters in diameter; tethers thicker than a few millimeters in diameter face a greater danger from orbital debris.

The 20-km-long, 0.75-mm-thick polyethylene fiber SEDS-2 tether, which was launched into a 540-km circular orbit by a Delta vehicle in 1994, was severed five days after launch, probably by a micrometeoroid.

removal to as much as 5 percent of the exposed paint areas (Coombs et al., 1992). Such loss of thermal control area is probably a minor issue, however, because it can be handled easily through oversizing in design. (Oversizing and other operational protection schemes are discussed in Chapter 6.) Perforation of thermal control blankets can also damage thermal control systems by delaminating layers and exposing protected components to the space environment (Allbrooks and Atkinson, 1992; Meshishnek et al., 1992).

Finally, small debris impacts can damage spacecraft solar power systems. Effects of debris impact can range from localized damage to cover glasses and solar cells to failure of strings of cells. Impacts can perforate or break exposed spacecraft cabling (including power cables), causing short circuits or failures. In addition, even small debris impacts can create plasmas, which can couple into solar arrays, causing failures (Krueger, 1993). Because of the phenomena associated with perforation through thin (compared to the impacting particle's diameter) films, however, the newer thin-film solar cell technologies are less susceptible to large-scale damage from small impacts.

FINDINGS

Finding 1: The probability that a spacecraft will be struck by debris is dependent on the spacecraft's orbital altitude and, to a lesser extent, its orbital inclination. The orbital regions where impact with medium or large debris is most likely are those between about 750- and 1,000-km altitude and those around 1,500-km altitude. Spacecraft in semi-synchronous orbits or GEO are, on average, probably about 100 times less likely to be struck by debris than most LEO spacecraft, and spacecraft not in any of the heavily used orbital regions (LEO, semi-synchronous orbit, or GEO) are even less likely to collide with debris.

Finding 2: Current models indicate that a collision in orbit will result in complete breakup if the ratio of the impactor's relative kinetic energy to the mass of the object with which it collides is greater than about 40 J/g. In LEO, debris as small as 0.1% of a space object's mass can cause the object to break up into many fragments. A typical LEO catastrophic collision involving a spacecraft may eject tens or hundreds of fragments large enough to cause a breakup if they collide with another spacecraft. At higher altitudes, where collision velocities are slower, a much larger impactor would be needed to cause catastrophic breakup.

Finding 3: Impacting space objects not large enough to break up a spacecraft can still cause significant damage through a variety of mechanisms, including perforation, spallation, and impulsive loading. The effect of debris impact on a particular spacecraft is strongly dependent on the spacecraft's design; an impact that could cause a poorly protected spacecraft to fail might do no damage to a well-protected spacecraft. Some spacecraft components (such as tethers) may, however, be very difficult to protect effectively.

Finding 4: Small debris impacts can degrade spacecraft surfaces and components. This degradation might have no effect on a spacecraft's capabilities, might reduce its functional lifetime, or might even cause the failure of components, depending on the impacted component and the energy of the impact. Although the mechanisms for some failures are obvious (e.g. a fluid leak caused by a hole or the consequences of a wire being severed), other damage-causing mechanisms and the associated effects on component performance are not well understood.

REFERENCES

Allbrooks, M., and D. Atkinson. 1992. The Magnitude of Impact Damage on LDEF Materials. NCR-188258. Houston, Texas: National Aeronautics and Space Administration Johnson Space Center.

Bernhard, R.P., T.H. See, and F. Horz. 1993. Projectile compositions and modal frequencies on the "Chemistry of Micrometeoroids" LDEF experiment. Pp. 551–573 in NASA Conference Publication 3194, LDEF–69 Months in Space: Proceedings of the Second LDEF Post-Retrieval Symposium, A.S. Levine, ed. Hampton, Virginia: NASA Langley Research Center.

Chobotov, V.A. 1991. The Space Debris Problem and Preliminary LDEF Results. M&D Sig NASA Model. Paper presented to the fourth ISCOPS Conference, Kyoto, Japan, November 17–20.

Coombs, C.R., D.R. Atkinson, M.K. Allbrooks, A.J. Watts, C.J. Hennessy, and J.D. Wagner. 1992. Damage areas on selected LDEF aluminum surfaces. P. 595 in NASA Conference Publication 3194, Part 2, Proceedings of the Second LDEF Post-Retrieval Symposium, A.S. Levine, ed. Hampton, Virginia: NASA Langley Research Center.

Cour-Palais, B.G. 1991. Meteoroid and Debris Shield Analysis. NASA Contract No. NAS 9-18200. Task order no. C91KC-033-R2. St. Louis, Missouri: McDonnell Douglas Space Systems.

Goldstein, R.M., and L.W. Randolph. 1992. Rings of Earth. IEEE Transactions on Microwave Theory and Techniques 40(6):1077–1080.

Grün, E., H.A. Zook, H. Fechtig, and R.H. Giese. 1985. Collisional balance of the meteoritic complex. Icarus 62:244–272.

Humes, D.H. 1994. Small craters on the meteoroid and space debris impact experiment. NASA Conference Publication 10120, Third LDEF Post-Retrieval Symposium Abstracts. Hampton, Virginia: NASA Langley Research Center.

Johnson, N.L. 1985. History of consequences of on-orbit break-ups. Advances in Space Research: Space Debris, Asteroids, and Satellite Orbits 5(2):11–19.

Kessler, D.J. 1969. Average relative velocity of sporadic meteoroids in interplanetary space. AIAA (American Institute of Aeronautics and Astronautics) Journal 7(12):2337–2338.

Kessler, D.J. 1972. A Guide to Using Meteoroid-Environment Models for Experiment and Spacecraft Design Applications. NASA TN D-6596, March. Houston, Texas: NASA Johnson Space Center.

Kessler, D.J. 1981. Derivation of the collision probability between orbiting objects: The lifetime of Jupiter's outer moons. Icarus 48:39–48.

Kessler, D.J. 1984. Tether Sever Rate from Meteoroids and Debris. NASA Memorandum SN3-84-217 August 21. Houston, Texas: National Aeronautics and Space Administration Johnson Space Center.

Kessler, D.J. 1993. Orbital debris environment. Pp. 251–262 in Proceedings of the First European Conference on Space Debris, Darmstadt, Germany, 5–7 April 1993. Darmstadt: European Space Operations Center.

Kessler, D.J., R.C. Reynolds, and P.D. Anz-Meador. 1989. Orbital Debris Environment for Spacecraft Designed to Operate in Low Earth Orbit. NASA TM 100-471. Houston, Texas: National Aeronautics and Space Administration Johnson Space Center.

Kessler, D.J., R.C. Reynolds, and P.D. Anz-Meador. 1994. Space Station Program Natural Environment Definition for Design, International Space Station Alpha. NASA SSP 30425, Revision B. Houston: National Aeronautics and Space Administration Space Station Program Office.

Krueger, F.R. 1993. Hypervelocity impact physics—Plasma discharge phenomena on solar generators. The Behavior of Systems in the Space Environment, NATO ASI Series E: Applied Sciences 245:273–290.

Love, S.G., and D.E. Brownlee. 1993. A direct measurement of the terrestrial mass accretion rate of cosmic dust. Science 262:550–553.

McKnight, D.S. 1993. Collision and Breakup Models: Pedigree, Regimes, and Validation/Verification. Briefing presented to the National Research Council Committee on Space Debris Workshop, Irvine, California, November 18.

Meshishnek, M.J., S.R. Gyetvay, K.W. Paschen, and J.M. Coggi. 1992. Long Duration Exposure Facility (LDEF) Experiment M0003 Meteoroid and Debris Survey. P. 357 in NASA Conference Publication 3194, Proceedings of the Second LDEF Post-Retrieval Symposium, A.S. Levine, ed. Hampton, Virginia: NASA Langley Research Center.

NASA (National Aeronautics and Space Administration. 1970). Meteoroid Damage Assessment. NASA Space Vehicle Design Criteria. NASA SP-8042. Washington, D.C.: NASA.

Simon, C.G., J.D. Mulholland, J.P. Oliver, W.J. Cooke, and P.C. Kassel Jr. 1993. Long-term microparticle flux variability indicated by comparison of interplanetary dust experiment (IDE) timed impacts for LDEF's first year in orbit with impact data for the entire 5.77-year orbital lifetime. Pp. 693-703 in NASA Conference Publication 3194, Proceedings of the Second LDEF Post-Retrieval Symposium, A.S. Levine, ed. Hampton, Virginia: NASA Langley Research Center.

Stansbery, E.G., D.J. Kessler, T.E. Tracy, M.J. Matney, and J.F. Stanley. 1994. Haystack Radar Measurements of the Orbital Debris Environment. JSC-26655 May 20. Houston, Texas: National Aeronautics and Space Administration Johnson Space Center.

Thompson, T.W., R.M. Goldstein, D.B. Campbell, E.G. Stansbery, and A.E. Potter, Jr. 1992. Radar detection of centimeter-sized orbital debris: Preliminary Arecibo observations at 12.5-cm wavelength. Geophysical Research Letters 19(3):257–259.

U.S. Space Command. 1994. U.S. Space Command Satellite Catalog. Cheyenne Mountain Air Force Base, Colorado: U.S. Space Command.

Watts, A.J., D.R. Atkinson, L. Crowell, B. Tritz, and S. Rieco. 1994. Impact Effects on Optics Survivability: Induced Optical Scatter. PL-TR-94-1046. Kirtland Air Force Base, New Mexico: Phillips Laboratory (in publication).

5

Tools for Damage Assessment and Prediction

There are three principal methods for assessing and predicting the damage caused to spacecraft by the impact of orbital debris: observation of such impacts in space, ground-based hypervelocity impact testing, and analytical or numerical (computer) simulation of impacts. Since it is very difficult to gather data from the rare impacts of medium-sized or large debris in space, assessment of the potential damage such debris can cause to space systems is accomplished primarily through experimental testing and analytic/numeric methods. Experimental testing generally provides the majority of information on these effects; analytic or numerical tools currently mainly supplement and extend experimental results.

GROUND-BASED HYPERVELOCITY TESTING

Experimental laboratory testing can simulate and/or verify three major types of orbital debris-related phenomena: (1) the effects of orbital debris impacts on spacecraft component performance, reliability, lifetime, and survivability; (2) the capabilities and performance of impact damage mitigation techniques, such as shielding and shuttering; and (3) the creation of orbital debris in collision-induced breakups of spacecraft and rocket bodies. The principal technique used to simulate these phenomena is hypervelocity impact testing.

Ground-based hypervelocity impact testing provides a means to determine how well various components, subsystems, or entire spacecraft will survive a collision with debris. Since it is infeasible to build entire spacecraft to be destroyed in ground tests, most tests are performed on components or on assemblies of components. Items tested can range from isolated fuel tanks and wiring harnesses to multicomponent assemblies including insulation materials and structural members (Christiansen, 1990; Christiansen and Ortega, 1990; Whitney, 1993; Schneider and Stilp, 1993). Although it is economically infeasible to test all components against all possible combinations of debris impact conditions, critical components can be evaluated with nominal impacts, and analytic or numerical techniques can then be used to extrapolate these results to other types of collisions.

Hypervelocity impact tests are also used to test and design debris shields. As with component testing, it is economically infeasible to test all possible shield configurations against all possible impact conditions, so a mixture of experimental testing, analytic methods, and numerical methods is used. Because the debris threat is not well enough known to "optimize" debris shielding against any particular type of impactor, shield designers develop shields to protect spacecraft against a wide range of impactor sizes, shapes, and velocities without greatly increasing the spacecraft's mass.

Finally, impact tests can be performed to examine the creation of fragmentation debris from breakups caused by hypervelocity collisions in space. This type of debris creation may play an important role in the evolution of the future debris population (as discussed in Chapter 8), but as mentioned in Chapter 2, only a few such tests have been performed to date. Such tests can be expensive, but since current data are very limited, a few well-planned and instrumented tests could add considerably to our knowledge of collision products and provide the basis for better estimates of the future debris population. Again, analytic and numerical methods can be used to extrapolate the limited test data to a wider range of possible situations.

The mass and velocity regimes required of an impactor in a hypervelocity test vary depending on the objective of the test. Obviously, the closer the tests come to matching real impactors' velocities, masses, materials, and shapes, the more accurate and useful the information acquired will be. For tests to determine the amount of debris created by a collision-induced breakup of a space object, it is necessary to use impactors large enough to fragment the target completely. For tests of spacecraft components and damage mitigation techniques, it is usually only necessary to use impactors that might feasibly be shielded against. The impactors used in such tests can range from millimeter to centimeter size, with masses ranging from much less than a gram up to several grams. Impactor shape must also be considered; since many potential debris impactors are fragments from rocket body or spacecraft explosions, the geometry of possible impacting objects can vary greatly. Finally, the tests must simulate the typical impact velocities for debris, which can range up to about 15 km/s in LEO and up to about 800 m/s in GEO.

Hypervelocity Test Capabilities

A wide range of experimental facilities have some capability to simulate orbital debris impact conditions. Table 5-1 summarizes the principal generally available hypervelocity impact facilities, and Figure 5-1 displays the capabilities of these facilities in terms of the projectile sizes they launch and the impact velocities that these projectiles reach. Figure 5-1 also points out size and velocity regimes of debris impacts that could potentially be shielded against but that cannot be achieved with current hypervelocity impact capabilities.

As seen in Figure 5-1, the capability exists to perform impact tests with even fairly large masses at velocities typical of collisions in high-altitude orbits. The U.S. military has conducted many tests (primarily for antiaircraft and armor/antiarmor purposes) in these mass and velocity

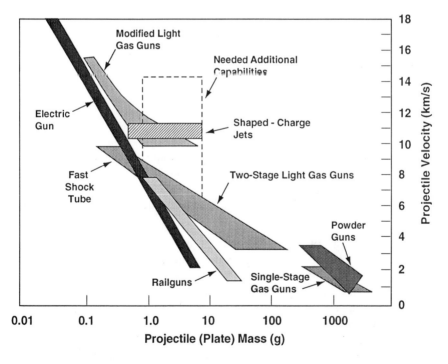

FIGURE 5-1 Capabilities of hypervelocity launch facilities.

TABLE 5-1 Experimental Test Facilities

	Maximum Mass (gm) at Maximum Velocity	Maximum Velocity (km/s)	Spheres	Rods
Single stage guns	2,000	1.4	X	X
Propellant guns	1,000	2.5	X	X
Two-stage light gas guns	0.01 to 250	8	X	X
Explosive Techniques				
Inhibited shaped charge	~1	~11		
Staged explosives	~0.1	14		
Advanced Hypervelocity Launchers				
Modified two-stage light gas gun (HVL)	0.2	15.8		
Fast shock tube	~1	~10		
Railgun	~1	~8		X
Other				
Electric discharge gun	0.01	~20	X	
Plasma drag accelerator	1E-07	~20	X	
Van de Graaff accelerator	1E-12	~100	X	

NOTE: Projectile masses and velocities are typical for the given capability. Increased mass capabilities have been achieved in modified scaled systems. "Other" refers to shapes that are not controlled (e.g., a slug in a jet tip). For more information on the velocity and size ranges achievable with various facilities, see Figure 5-1.

regimes. For example, some work done on "hit-to-kill" missile impacts could possibly be applied to collisions between large bodies at a few kilometers per second. However, the applicability of these data to orbital debris issues has not been studied, and in any case, the data may be considered too sensitive for wide release.

For studies of debris impacts at higher velocities, the standard laboratory tool is the two-stage light gas gun. Conventional light gas guns come in a variety of sizes and typically can accelerate impacting objects from less than 1.5 mm to more than 50 mm in diameter to about 8 km/s,

Plates	Other	Comments
X		Typically used for material property studies. Typical projectile diameter is 100 mm.
X		Typically used for material property studies. Typical projectile diameter is 90 mm.
X		Maximum size is customized for each gun. Microgram spheres have been launched to about 10.5 km/s using special designs.
	X (hollow cylinders)	Projectile length to diameter difficult to control.
X		Computer analysis required to establish thermodynamic state. Typically used for material property studies.
X	X	Computer analysis necessary to characterize thermodynamic state of the projectile. Photography necessary to characterize projectile shape.
X	X	Still in development. Different variations claim capability to launch gram-size projectiles to several tens of kilometers per second or 100 projectiles to about 10 km/s. Still in development stage. Plastic rods have been launched to velocities approaching 8 km/s.
		Thermodynamic state of projectile not well characterized. Thermodynamic state of projectile not well characterized. Thermodynamic state of projectile not well characterized.

although some facilities have been able to accelerate smaller particles to 10 km/s or more (often damaging the gun in the process). The standard projectile used in a two-stage light gas gun is a sphere, although various other shapes, including thin plates, cylinders, and long rods, can be launched (Piekutowski, 1986).

Light gas guns cannot launch impactors to the velocities typical of LEO debris impacts (10-15 km/s), but several ultrahigh-speed launchers have been developed that extend the impact velocity range for debris impact studies. One is a modified light gas gun technique, referred to as the Hypervelocity Launcher (Chhabildas et al., 1992a). This technique recently launched 1-mm thick, 6-mm diameter titanium plates to velocities of 15.8 km/s (Chhabildas, in press). Similarly modified larger light gas guns have launched 2-mm-thick, 30-mm-diameter titanium plates to velocities greater than 10 km/s. This resultant capability covers the velocity and mass regime of a large fraction of space debris (see Figure 5-1) but is limited to plate-shaped impactors, and numerical methods are needed to specify the thermodynamic state of the impacting particle. A technique with the potential to further extend the range of debris impact studies is the electromagnetic railgun (Asay et al., 1989), which may eventually have the capability of launching spheres up to a centimeter in diameter to velocities of 15 km/s.

Launchers using explosives also have potential use in debris studies. One such technique employs an inhibited shaped-charge explosive to launch objects with dimensions in the tens of millimeters to velocities of about 11 km/s (Walker et al., 1992). Velocities relevant to space debris studies are therefore realizable, but the objects launched by such explosives are typically hollow cylinders with varying length-to-diameter ratios. This unusual shape complicates analysis of the data because analytical models for the damage caused by objects of this shape have not yet been developed. NASA is using a light gas gun to launch hollow cylinders at velocities of up to 8 km/s in order to learn more about the damage caused by this type of projectile. Russian investigators (Isbell et al., 1992) used a different staged-explosive technique to launch thin plates at high velocities for equation-of-state studies. This method has launched flat plates with centimeter diameters to velocities greater than 15 km/s. Finally, Russian and American investigators are developing a fast shock-tube device with the potential to launch larger flat plates to velocities of 15 km/s.

Other advanced launcher techniques have been developed to extend the range of small particle impacts to even higher velocities. Plasma drag launchers can launch microgram particles to velocities of nearly 20 km/s, and electrostatic launchers have extended this range to more than 100 km/s for particle masses of 10^{-15} gram (Stradling et al., 1992). Capacitor discharge techniques can launch thin flyers of metal and plastic with masses of a few tens of milligrams to velocities of 20 km/s (Lee et al., 1992). These techniques are most commonly used to simulate the damage caused to particular components by the impact of small debris and micrometeoroid particles. They have not, however, been widely used to develop damage prediction or degradation models; such models are based primarily on data analysis of returned spacecraft surfaces.

Techniques also have been developed to simulate high-velocity debris impacts without actually launching impactors at orbital velocities. One such technique, developed to overcome the velocity and mass limitations of existing hypervelocity test facilities, is "dissimilar materials"

BOX 5-1   Two-Stage Light Gas Guns

Two-stage light gas guns pressurize and accelerate gas to launch small projectiles to high velocities. A first-stage launch tube, typically 80-100 mm in diameter, contains a low molecular weight gas (such as hydrogen or helium) pressurized to a few atmospheres. An explosive charge accelerates a heavy piston to about 0.5 km/s, which accelerates the gas through a conical section to a smaller diameter launch tube (the second stage) containing the projectile. The accelerated gas produces a loading pressure of about 10,000 atmospheres, which accelerates the projectile to velocities of 8-10 km/s. The projectile then strikes the target in an evacuated chamber.

In the Hypervelocity Launcher, the second-stage projectile impacts a thin plate at the end of the launch tube. The projectile is designed to impact so as to create a nearly isotropic pressure loading, accelerating the plate to velocities of up to 15 km/s, depending on its mass.

---

testing (Holsapple, 1992). This method simulates the impact of aluminum particles on aluminum plates at velocities exceeding 12 km/s (where vaporization of the impacted materials occurs) by impacting cadmium (or another low-melting-point material) particles on cadmium plates at velocities of about 6 km/s. This method may be useful in investigating aspects of hypervelocity impact phenomena encountered during high-speed impact on aluminum, but determining the extent to which the results of dissimilar materials testing are applicable to damage prediction at velocities of about 10 km/s requires further detailed investigation and evaluation.

Russian investigators have developed a more radical method to simulate target conditions produced by ultrahigh-velocity particles; rather than launching an impactor, they have used electron beams and laser deposition to simulate the kinetic energy of high-velocity particles (Anisimov et al., 1985). Researchers in the United States, Germany, and Israel have also done extensive work on simulating impacts using ion beams and lasers (Gilath et al., 1992; Krueger, 1993) and have worked on laser ablation techniques for accelerating particles to high velocities (Trott and Meeks, 1990). To accurately simulate high-velocity impact, such techniques must not only match the impacting particle's kinetic energy but also impart such energy over a similar time frame to that of an impact and account for such effects of impact as momentum transfer and changes of state. Currently, these experiments cannot achieve such an accurate simulation of an impact, so analytic and numerical methods are used to convert their data into damage assessment predictions.

Sharing Hypervelocity Impact Information

One of the main reasons for the lack of good models of hypervelocity impact damage is that hypervelocity test data are not formally shared, and the capabilities of many facilities involved in hypervelocity testing are not commonly known. The general inaccessibility of facility capabilities and impact data generated at these facilities has resulted in duplication of effort, expense, and delays. Although information about the capabilities of laboratory facilities able to study debris impacts can usually be obtained from a variety of sources, such as published journals, company brochures, and word of mouth, there is no systematic process for obtaining this information. Detailed information regarding the capabilities of a specific laboratory is usually acquired through individual visits by researchers. Often, this information is published in trip reports and other company documentation and is not widely disseminated. This is especially true of facility capabilities outside the United States and of "alternative" techniques for simulating debris impact conditions, such as the laser and electron-beam facilities that Russian investigators use.

Attempts to disseminate information about the capabilities of hypervelocity impact facilities have been made; information on U.S. impact testing capabilities was at one point compiled in the "Facilities Handbook" (Malley and Nicols, 1987). The goal of this handbook was to (1) determine where impact testing could be conducted; (2) identify "holes" in test capabilities, facilities, and instrumentation; and (3) provide a mechanism to identify the most effective test facilities and methods for filling these holes. There was a great deal of useful information in this report, but it did not cover all U.S. facilities and covered none outside the United States. In addition, it did not include "alternative" test facilities, such as shaped charges, Van de Graaff accelerators, and electron beam deposition. Finally, distribution was limited to U.S. Air Force facilities, and the information was not entered into a database for easy retrieval.

Even more pressing than the lack of information about facility capabilities is the general inaccessibility of debris impact data generated at various facilities. Many test facilities have extensive collections of data (sometimes going back 30 to 40 years), most of which are not computerized or stored in databases for easy access. Often, these data are published in company reports that have limited distribution and are not archived for public access. In addition, technical information from many countries is published in journals that are not easily accessible in other nations or, in some cases, is not published at all because of potential or past military secrecy constraints. This inaccessibility of a great deal of data has surely limited the development of good models of debris impact damage.

ANALYTICAL AND NUMERICAL MODELING
OF DEBRIS IMPACTS

Analytic/numeric methods of various levels of complexity are used to predict the response of spacecraft to hypervelocity debris impacts. Analytic methods developed to aid spacecraft designers in the design of protective shields are the least complex. These include (1) "ballistic limit" equations (Cour-Palais, 1986; Herrman and Wilbeck, 1986; Reimerdes et al., 1993), which calculate the size of a particle stopped by a particular shield as a function of impact speed, impact angle, and impactor density; and (2) shield sizing equations (Christiansen, 1993), which provide estimates of shield thicknesses and spacings required to protect against particles of given sizes, velocities, densities, and impact angles. Shield sizing equations may incorporate ballistic limit equations to determine the effects of impact on the individual walls that make up the shield.

Analytic methods available to spacecraft designers for predicting the damage caused by impacts, and the effects of that damage, are slightly more complex. These include (1) impact damage and effects equations derived from physical principles (Watts et al., 1993, Watts et al., in press) and (2) semiempirical impact cratering, perforation, and spallation equations (e.g., Cour-Palais, 1979; Carey et al., 1984; Hörz et al., 1994). Other analytic models that are useful for providing a qualitative understanding of impact damage include the Grady model (Grady, 1987; Grady and Passman, 1990), the Tate model (Tate, 1967, 1969), and the Ravid and Bonder model (Ravid and Bonder, 1983; O'Donoghue et al., 1989).

There are, however, currently no standardized risk assessment models to determine the probability of component degradation or failure due to orbital debris impacts. Performance degradation models are also not standardized and currently exist for only a few component types. Because of this, spacecraft degradation due to debris impact is currently modeled by combining basic engineering model predictions of the expected environment with empirical scaling models for damage prediction. These empirical scaling laws, though, must often be applied via unproven extrapolations to materials and velocities that were not included in the original data sets on which the empirical models were based. After predicting damage, simple performance degradation "rules" relating degradation to damage area can be applied to determine whether performance will remain within specifications.

Empirical equations based on ballistic limit curves or other experimental data are often used to predict the performance of debris shields. These equations can produce good results if experimental data have been generated for similar particle configurations and velocities (Christiansen,

1993). A current deficiency of these methods is their typically being based on empirical data obtained over limited velocity and impactor shape regimes. This produces considerable uncertainty in extrapolating these equations to other materials, higher velocities, or velocity regimes where phase changes (such as vaporization) occur.

Numerical simulations also can be used to predict the damage to spacecraft from debris impacts or to determine the characteristics of the fragmentation debris released in spacecraft or rocket body breakups. Some such computer codes, usually referred to as "hydrocodes," can model the spacecraft and impact in three dimensions, though many calculations are performed in two dimensions, particularly when the code is being used for "phenomena scoping" and "parameter sensitivity" calculations (i.e., to determine the degree to which changes in material properties would change the size or shape of the impact damage).

The accuracy of results derived from these codes depends on the resolution with which the components are modeled and the material models used in the computations. Good models of the properties of materials and equations of state do not exist for many of the newer materials used on spacecraft, including many composites, ceramics, and coatings. If such models are not developed, these codes may have limited future value. Good material models must also accurately represent phase changes caused by the impact (such as vaporization) as well as material strength effects (such as compressive and tensile failure behavior).

The memory and speed of available computers limited the numerical resolution of early computer simulations of debris damage. Recent developments in computer capabilities have mitigated these problems; it is now possible to model individual components, such as debris shields and hull plates, with sufficient numerical resolution to predict debris impact damage (if good material models of all components are used) with reasonably good accuracy (Hertel et al., 1992; Hertel, 1993; Farenthold, 1992; and Katayama et al., 1993).

Computer simulations are most reliable, however, when benchmarked against experimental data obtained with materials, particle shapes, and velocity regimes similar to those being simulated (Chhabildas et al., 1992b) and used to interpolate between good experimental data. Often, though, experimental data are not available, so numerical analyses provide the only information available for specific impact conditions. In these cases, predictions of debris impact damage must be used with caution. When combined, hypervelocity testing and computer modeling are powerful tools for assessing the survivability of space systems to debris impacts.

LIMITATIONS IN DAMAGE ASSESSMENT AND PREDICTION CAPABILITIES

As Figure 5-1 and Table 5-1 show, the range of capabilities for launching particles of the correct mass, velocity, and shape to simulate space debris impacts is limited. This has led to some limitations in current damage assessment and prediction capabilities that have serious implications for the debris field. These are (1) that the full variety of debris shapes and compositions likely to exist in orbit cannot yet be tested in all velocity regimes, and (2) that there is difficulty in launching larger impactors to typical LEO collision velocities. The first limitation makes shield design against the actual debris environment difficult. The second limitation not only reduces the accuracy of damage predictions for the impact of centimeter-size objects, but also contributes to the uncertainty in predictions of the future debris population.

Many analytic theories and measures of impact damage, such as the ballistic limit, are based on the impact of spherical particles. While this is a reasonable assumption for meteoroid impacts, space debris exhibit a much wider assortment of shapes. It has been known for some time that nonspherical impactors can do more damage than spherical impactors in many situations. For example, penetration depth and crater volume from impacts in thick plate targets are strongly influenced by the length of the projectile along its flight axis (Gehring, 1970). Figure 5-2 illustrates how crater depth and volume in a thick target can vary by impactor shape. For Whipple bumper shields (described in Chapter 6), flat plate projectiles are generally more damaging than spherical projectiles of the same mass and velocity (Boslough et al., 1993). Figure 5-3 illustrates how the size of the rupture on the backwall of a Whipple bumper shield can vary greatly with impactor shape. Because of these shape effects, shields designed based on experience with spherical impactors may not be as effective as predicted in protecting spacecraft from actual orbital debris impacts.

Another weak link in current meteoroid and debris shield development efforts is that, because of the limited data available regarding the distribution of material types in the debris environment, models used for shielding design generally assume that large objects are composed of aluminum and small objects are composed of aluminum oxide. Some debris, however, is composed of higher-density materials; LDEF detected impacts by stainless steel, copper, and silver particles (Hörz and Bernhard, 1992). This is a problem because a shield that is designed to withstand only aluminum projectiles could potentially be perforated by high-density debris or meteoroids.

It is not feasible, however, to solve these problems by testing shields

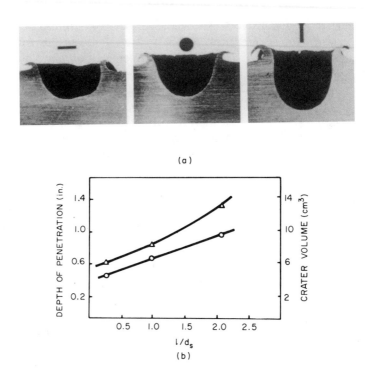

(a)

(b)

FIGURE 5-2 Variation of crater size with impactor shape for a thick target.
SOURCE: Gehring, 1970.

and other components against all possible debris shapes and compositions (and sizes, masses, or velocities). Instead, analytic and numerical
methods can be used to extend a limited set of experimental results to
other configurations, shapes, compositions, etc., to identify worst-case
conditions that can be used in the design of spacecraft protection systems. If these computer simulations are validated with sufficient experimental data, reasonable confidence could then be assigned to the com-
puted results. This approach could increase the reliability of a given
protection system and minimize the possibility of serious over- or under-
design.

The inability to launch large impactors at typical LEO collision velocities not only causes the same type of problems described above but
also limits the accuracy of breakup models. Currently, masses capable of
breaking up even the smallest spacecraft can be launched only to low

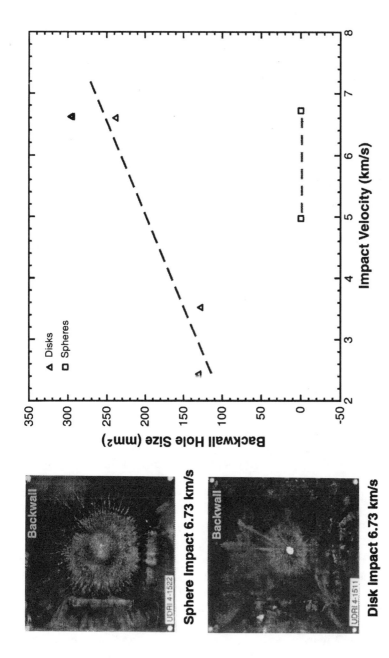

FIGURE 5-3 Whipple bumper backwall hole size as a function of impact velocity and impactor shape.
SOURCE: Konrad et al., 1993.

velocities (<7 km/s). Because of this lack of capability, all breakup models use data from tests at impact velocities lower than the typical LEO collision velocities. In addition, few facilities can perform large-scale collision tests in a controlled environment. (A very large chamber capable of conducting both explosion- and collision-induced breakup experiments has, however, been constructed in Russia [Fortov, 1993].) As mentioned in Chapter 2, without a controlled environment, data on the distribution of small particles generated by a breakup are suspect and data on the breakup-induced velocities of any size particle become difficult to obtain.

FINDINGS

Finding 1: High confidence in the validity of (1) assessments of the response of spacecraft components and shield configurations to debris impacts and (2) component and shield qualification and acceptance tests can presently be provided only by hypervelocity impact testing, but current hypervelocity impact facilities cannot simulate the full range of debris impactor sizes, compositions, shapes, and velocities. As a result, spacecraft protection systems currently are designed to resist the type of projectiles that can be launched by these facilities (most typically aluminum spheres). Because actual debris objects typically have more complex shapes that are very likely to do more damage than spheres at LEO collision velocities, current spacecraft surfaces and shield designs may not provide the desired level of protection.

Finding 2: Facilities in a number of nations are capable of carrying out hypervelocity impact tests for debris research but information about and access to these facilities is often difficult to obtain, there is no coordinated interfacility approach to either impact research or new facility development, and the results of experiments are not widely available. The general inaccessibility of facility capabilities and of the impact data generated at these facilities has resulted in considerable duplication of effort, slowing the development of good models of debris impact damage.

Finding 3: Analytical models can be used to design spacecraft shielding and to predict impact damage for regimes that hypervelocity testing cannot simulate. Numerical simulations can be useful tools for predicting damage to spacecraft and determining the characteristics of breakup debris. Unless both of these methods are validated by comparison to experimental data, however, significant variability in predicted results can occur. When used together, hypervelocity testing and computer modeling are powerful tools for assessing the survivability of space systems to debris impacts.

REFERENCES

Anisimov, S.I., B.A. Demidov, L.I. Rudakow, R.Z. Sagdeyev, and V.Ye. Fortov. 1985. Modeling the destruction of protective screens of the Vega spacecraft with the aid of high-current relativistic electron beams. Journal of Experimental and Theoretical Physics Letters 41(11):455.

Asay, J.R., T.G. Trucano, and R.S. Hawke. 1989. The use of hypervelocity launchers to explore previously inaccessible states of matter. International Journal of Impact Engineering 10:51–66.

Boslough, M.B., J.M. Ang, L.C. Chhabildas, W.D. Reinhardt, C.A. Hall, B.G. Cour-Palais, E.L. Christiansen, and J.L. Crews. 1993. Hypervelocity testing of advanced shielding concepts for spacecraft against impacts to 10 km/s. International Journal of Impact Engineering 14:95–106.

Carey, W.C., J.A.M. McDonnell, and D.G. Dixon. 1984. An empirical penetration equation for thin metallic films used in capture cell techniques. Pp. 131–136 in Properties and Interactions of Interplanetary Dust. Proceedings of the 85th Colloquium of the International Astronomical Union, Marseille, France, July 9–12. Volume 19 of the Astrophysics and Space Science Library. Dordecht, Holland: D. Reidel Publishing Company.

Chhabildas, L.C. In press. To be published in Proceedings of the 1994 Hypervelocity Impact Symposium, Santa Fe, New Mexico, October 16–19. International Journal of Impact Engineering.

Chhabildas, L.C., J.E. Dunn, W.D. Reinhart, and J.M. Miller. 1992a. An impact technique to accelerate flier plates to over 12 km/s. International Journal of Impact Engineering 14:121–132.

Chhabildas, L.C., E.S. Hertel, and S.A. Hill. 1992b. Hypervelocity impact tests and simulations of single Whipple bumper shield concepts at 10 km/s. International Journal of Impact Engineering 14:133–144.

Christiansen, E.L. 1990. Investigation of hypervelocity impact damage to space station truss tubes. International Journal of Impact Engineering 10:125–133.

Christiansen, E.L. 1993. Design and performance equations for advanced meteoroid and debris shields. International Journal of Impact Engineering 14:145–156.

Christiansen, E.L.,and J. Ortega. 1990. Hypervelocity Impact Testing of Shuttle Orbiter Thermal Protection System Tiles. AIAA Paper No. 90-3666. American Institute of Aeronautics and Astronautics Space Programs and Technologies Conference, Huntsville, Alabama, September 25–28. Washington, D.C.: American Institute of Aeronautics and Astronautics.

Christiansen, E.L., R. Bernhard, J. Hyde, J. Kerr, K. Edelstein, J. Ortega, and J. Crews. 1993. Assessment of high velocity impacts on exposed shuttle surfaces. Pp. 447–452 in Proceedings of the First European Conference on Space Debris, Darmstadt, Germany, 5–7 April 1993. Darmstadt: European Space Operations Center.

Cour-Palais, B.G. 1979. Space Vehicle Meteoroid Shielding Design. ESA SP-153. Paris: European Space Agency.

Cour-Palais, B.G. 1986. Hypervelocity impacts in metals. International Journal of Impact Engineering 5:221–238.

Farenthold, E.P. 1992. A Lagrangian model for debris cloud dynamics simulation. International Journal of Impact Engineering 14:229–240.

Fortov, V. 1993. Hypervelocity Impact in Space Debris Problem. Briefing presented to the Committee on Space Debris Workshop, National Research Council, Irvine, California, November 18.

Gehring, J.W. Jr. 1970. Engineering considerations in hypervelocity impact. Pp. 466–468 in High-Velocity Impact Phenomena, R. Kinslow, ed. New York-London: Academic Press.

Gilath, I., S. Eliezer, T. Bar-Noy, R. Englman, and Z. Jaeger. 1992. Material response at hypervelocity impact conditions using laser induced shock waves. International Journal of Impact Engineering 14:279.

Grady, D.E. 1987. Fragmentation of rapidly expanding jets and sheets. International Journal of Impact Engineering 5:285–292.

Grady, D.E., and S.L. Passman. 1990. Stability and fragmentation of ejecta in hypervelocity impact. International Journal of Impact Engineering 10:192–212.

Herrman, W., and J.S. Wilbeck. 1986. Review of hypervelocity penetration theories. International Journal of Impact Engineering 5:307–322.

Hertel, E.S. 1993. Simulation of orbital debris impacts on bumper shields. Pp. 413–418 in Proceedings of the First European Conference on Space Debris, Darmstadt, Germany, 5–7 April 1993. Darmstadt: European Space Operations Center.

Hertel, E.S., L.C. Chhabildas, L. Yarrington, and S.A. Hill. 1992. Computational determination of ballistic limits for a simple Whipple shield. Pp. 15–18 in Hypervelocity Impacts in Space, J.A.M. McDonnell, ed. Canterbury, United Kingdom: University of Kent.

Holsapple, K.A. 1992. Hypervelocity impact experiments in surrogate materials. International Journal of Impact Engineering 14:335–345.

Hörz, F., and R.P. Bernhard. 1992. Compositional Analysis and Classification of Projectile Residues in LDEF Impact Craters. NASA TM-104750. Houston, Texas: National Aeronautics and Space Administration Johnson Space Center.

Hörz, F., M. Cintala, R.P. Bernhard, and T.H. See. 1994. Dimensionally scaled penetration experiments: Aluminum targets and glass projectiles 50 μm to 3.2 mm in diameter. International Journal of Impact Engineering 15(3):257–280.

Isbell, W.M., C.E. Anderson, J.R. Asay, S.J. Bless, D.E. Grady, and J. Sternberg. 1992. Penetration Mechanics Research in the former Soviet Union. Unpublished Foreign Applied Sciences Assessment Center Technical Assessment Report. September. McLean, Virginia: Science Applications International Corporation.

Katayama, M., T. Aizawa, S. Kibe, and S. Toda. 1993. A numerical simulation of hypervelocity impact of space debris against the Whipple bumper system. Pp. 419–424 in Proceedings of the First European Conference on Space Debris, Darmstadt, Germany, 5–7 April 1993. Darmstadt: European Space Operations Center.

Konrad, C.H., L.C. Chhabildas, M.B. Boslough, A.J. Piekutowski, K. Poorman, S.A. Mullin, and D.L. Littlefield. 1993. Dependence of debris cloud on projectile shape. Pp. 1845–1848 in High Pressure Science and Technology, S.C. Schmidt, J.W. Shaner, G.A. Samara, and M. Ross, eds. Woodbury, New York: American Institute of Physics.

Krueger, F.R. 1993. Hypervelocity impact physics—Plasma discharge phenomena on solar generators. The Behavior of Systems in the Space Environment, NATO ASI Series E: Applied Sciences 245:273–290.

Lee, R., J. Osher, H. Chau, M. Gerassimenko, G. Pomykal, and R. Spear. 1992. The impact of flat, thin plates on aluminum targets in the 5-10 km/s velocity range. International Journal of Impact Engineering 14:451–466.

Malley, M.M., and W.G. Nicols. 1987. Survey of Laser and Hypervelocity Test Facilities for the Defensive Shields Demonstration. Air Force Weapons Laboratory report AFWL-NTC-ITR-87-02 June (unpublished, distribution limited to Air Force Weapons Laboratories only).

O'Donoghue, P.E., S.R. Bonder, C.E. Anderson, Jr., and M. Ravid. 1989. Comparison of a high velocity impact model with numerical simulation. International Journal of Impact Engineering 8(4):289–301.

Piekutowski, A.J. 1986. Debris clouds generated by hypervelocity impact of cylindrical projectiles with thin aluminum plates. International Journal of Impact Engineering 5:509–518.

Ravid, M., and S.R. Bonder. 1983. Dynamic perforation of viscoplastic plates by rigid projectiles. International Journal of Engineering Science 21(6):577–591.

Reimerdes, H., K. Stecher, and M. Lambert. 1993. Ballistic limit equations for the Columbus double-bumper shield concept. Pp. 433–439 in Proceedings of the First European Conference on Space Debris, Darmstadt, Germany, 5-7 April 1993. Darmstadt: European Space Operations Center.

Schneider, E., and A. Stilp. 1993. Meteoroid/debris simulation at Ernst-Mach-Institut (EMI): Experimental methods and recent results. Pp. 401–404 in Proceedings of the First European Conference on Space Debris, Darmstadt, Germany, 5–7 April 1993. Darmstadt: European Space Operations Center.

Stradling, G.L., G.C. Idxorek, B.P. Shafer, H.L. Curling, Jr., M.T. Collopy, A.A. Hopkins Blossom, and S. Furerstenau. 1992. Ultra-high velocity impacts: Cratering studies of microscopic impacts from 3 km/s to 30 km/s. International Journal of Impact Engineering 14:719–728.

Tate, A. 1967. Theory for the deceleration of long rods after impact. Journal of Mechanical Physics Solids 15:387–399.

Tate, A. 1969. Further results in the theory of long rod penetration. Journal of Mechanics Physics Solids 17:141–150.

Trott, W.M., and K.D. Meeks. 1990. Acceleration of thin foil targets using fiber-coupled optical pulses. Pp. 997–1000 in Shock Compression of Condensed Matter—1989. Amsterdam: Elsevier Science Publishers.

Walker, J.D., D.J. Grosch, and S.A. Mullin. 1992. A hypervelocity fragment launcher based on an inhibited shaped charge. International Journal of Impact Engineering 14:763–774.

Watts, A., D. Atkinson, and S. Rieco. 1993. Dimensional Scaling for Impact Cratering and Perforation. NASA NCR-188259. March 16, Houston, Texas: National Aeronautics and Space Administration Johnson Space Center.

Watts, A.J., D.R. Atkinson, L. Crowell, B. Tritz, and S. Rieco. In press. Impact Effects on Optics Survivability: Induced Optical Scatter. PL-TR-94-1046. Kirkland Air Force Base, New Mexico: Phillips Laboratory.

Whitney, J.P. 1993. Hypervelocity Impact Tests of Shielded and Unshielded Pressure Vessels. JSC-32294. Houston, Texas: National Aeronautics and Space Administration Johnson Space Center.

6

Designing for the Debris Environment

As shown in Chapter 4, orbital debris poses a potential hazard to spacecraft in Earth orbit. Although a few measures to reduce the creation of new debris already have been taken, and it appears likely that more will follow, these efforts generally aim at averting major increases in—rather than actually reducing—the future debris population. Therefore, the only foreseeable significant reductions in the debris population will be those caused by orbital decay. The result is that even if measures are taken to minimize the creation of new debris, a debris hazard to spacecraft will exist for many years in most orbits. If measures to reduce the creation of new debris are *not* taken, the debris hazard in many orbits will increase (as discussed in Chapter 8). In any case, spacecraft designers and operators will have to deal with a debris hazard far into the future.

In the past, most spacecraft designers did not consider the debris hazard as a design consideration, due perhaps to a general lack of awareness of the threat, the low level of the perceived hazard, or an unwillingness to undertake the seemingly large research task of quantifying the risk and determining appropriate means to protect their spacecraft. Although large uncertainties still remain, an improved understanding of the debris environment, combined with the growing availability of analytic and experimental tools to quantify the threat to a spacecraft from debris and the development of techniques to protect against debris impacts, now makes it feasible for designers to assess the debris hazard and protect their spacecraft appropriately.

For spacecraft designers and operators, the decision whether and how

119

to protect their spacecraft against debris impact must involve balancing the risk and cost of damage from debris impact against the expense of implementing measures to protect against debris. The final decision will be different for each spacecraft because the hazard, acceptable risk, and cost of protection will vary depending on the spacecraft's orbit, configuration, and particular mission. One factor that will not vary greatly is that the earlier debris considerations are factored into the design process, the less costly will any necessary modifications be. Early in the design process, designers can modify aspects of the design to meet debris-related requirements at a minimum cost; later, however, the many design choices that have already been made and cannot easily be changed constrain further design changes.

Determining the need for (and extent of) protective measures against debris is a three-step process. First, the hazard from debris must be calculated by determining the size-dependent debris flux that the spacecraft is likely to experience and then determining the probability that the flux will damage the spacecraft. Second, the effectiveness of various protective methods (such as shielding or component rearrangement) that could be used to reduce the hazard must be determined. The final step is to look at the results of these two analyses and consider the tolerable level of hazard for the spacecraft, to determine the costs and benefits of implementing protective measures. As illustrated in Box 6-1, the final decision on protecting a spacecraft will vary greatly depending on the spacecraft involved and the level of hazard acceptable to the designers and operators.

DETERMINING THE HAZARD FROM DEBRIS

To quantify the threat of orbital debris to a spacecraft, designers must analyze the particular debris environment in their spacecraft's orbit, as well as the spacecraft's vulnerability to that environment. A number of analytic and experimental tools that can be very helpful in carrying out these tasks are now available to designers. It is important, though, that spacecraft designers who use these tools recognize the assumptions incorporated in them so that they fully understand the uncertainties associated with their output.

The overall flux of orbital debris that a spacecraft will experience is largely a function of the spacecraft's size, orbital altitude, inclination, and attitude; the duration of the mission; and the current level of solar activity. As discussed in Chapter 2, a number of orbital debris environmental models that designers can use to estimate the debris flux on spacecraft have been created. One detailed engineering model has been developed by scientists at NASA (Kessler et al., 1989) and is being used by NASA,

---

BOX 6-1 Design Responses to the Debris Hazard: Three Examples

International Space Station: Because of the extremely high value of the spacecraft and the desire to protect the people that will inhabit it, the International Space Station design requirements are that the probability that debris impact will cause a critical failure must be less than 0.5% per year.

Geostationary Communications Spacecraft: Due to the low perceived hazard in the geostationary orbit, no spacecraft in GEO are known to have design requirements specifically for protection against debris impacts, though they are designed to survive the micrometeoroid environment.

RADARSAT: The RADARSAT spacecraft is designed to be launched into an orbital regime with a high debris flux. The response of the RADARSAT designers is presented in some detail at the end of this chapter.

ESA, the National Space Development Agency of Japan, and the Russian Space Agency in the design of the International Space Station (Kessler et al., 1994). A simplified version of this model, accessible on the EnviroNET database (Lauriente and Hoegy, 1990), can predict the cumulative debris flux of a given size on a spacecraft surface in any LEO. The ESA Reference Model for Space Debris and Meteoroids is also available in an analytic form useful for spacecraft designers (Sdunnus and Klinkrad, 1993).

Once the debris flux and the distribution of impact angles have been estimated, the number of impacts on specific spacecraft components can be predicted. This process involves determining the location of each component relative to all the others and to the incoming space debris, to see how components shield one another and to determine where and at what angles debris is likely to strike each component. NASA's BUMPER probability analysis code (Christiansen, 1993), which was developed for the analysis of Space Station Freedom and has since been applied to the U.S. shuttle orbiter, LDEF, Mir, and the proposed International Space Station, can be used to link the debris (and meteoroid) environment with the spacecraft's geometry and penetration equations to determine the perforation hazard to each part of the spacecraft and to size shielding to prevent such perforations. However, BUMPER can only predict perforation hole size; it cannot predict other types of impact damage.

Other models, analyses, or impact tests are needed to assess the probability of component failures due to impact damage effects. As described in Chapter 5, this can be accomplished through numerical or analytical methods, by subjecting some components to actual hypervelocity impacts, or through a combination of both approaches. As described in

Chapter 4, several different impact effects should be considered when making such assessments. These include

- the effect of perforations on the overall performance of the system;
- damage resulting from high-velocity fragments, plasmas, and impulsive loads generated by the debris impact;
  - the extent and effect of surface degradation from debris impact;
  - the growth of impact damage features over time; and
  - damage to critical components leading to spacecraft loss.

The performance of shielding and operational protection techniques in preventing these types of damage from debris impact can also be explored.

Finally, the vulnerability of the spacecraft to debris can be determined by combining the probability of failure of its various components due to debris impact. This includes accounting for the redundancy of components and their criticality to the spacecraft. If the vulnerability of the spacecraft is found to be unacceptable, various protective measures can be taken to decrease the threat to the spacecraft as a whole, or at least to protect its more vulnerable components.

DAMAGE PROTECTION TECHNIQUES

Passive, active, or operational protection schemes can be used to protect spacecraft from debris impact damage. *Passive protection* generally consists of shielding a spacecraft or its critical components. *Active protection* schemes use sensors to provide advanced warning of impact and then protect critical components or move the spacecraft to avoid the potential impact. *Operational protection* schemes change the design of a spacecraft to allow for graceful degradation or change a spacecraft's operations to reduce the overall hazard to the mission. Designers who wish to protect their spacecraft from debris impact must trade off the costs and the benefits of each method to determine the appropriate method or methods with which to protect the spacecraft.

Passive Protection

Passive protection typically involves the shielding of a spacecraft against debris impact. As a result of the size distribution of objects in Earth orbit (as illustrated in Figure 4-1), spacecraft are much more likely to be struck by small debris than by medium-sized debris; the chance of being struck by large debris is lower still. For obvious reasons, the mass of shielding needed to protect a spacecraft against larger, more energetic objects increases with the objects' size; this growth in shield mass will increase a spacecraft's launch costs or decrease its payload mass. The decision to shield a spacecraft and the determination of how much shielding is necessary require that the acceptable level of risk (i.e., the probability of collision with an object large enough to damage the spacecraft) be balanced against the added mass (and thus cost) required to protect the spacecraft against various debris size ranges.

In practice, the basic spacecraft structure, which must be more massive than is needed in space simply to withstand launch loads, often becomes the primary "shield" against debris. Only if this structure is incapable of providing sufficient protection should additional shielding be considered. If additional shielding of components is required, existing components on spacecraft often can be augmented to serve as debris shields. For example, component walls can be thickened or layers of particle-breaking material can be added to thermal blankets covering the exterior of a spacecraft. Although this type of modification will not provide as much protection as the equivalent mass of specially designed shields, it generally results in smaller increases in spacecraft volume, complexity, and cost.

If specially designed shields are deemed necessary, the driving issue is to minimize mass, size, and cost, while maximizing protection against debris impact damage. Two basic types of shields, monolithic and spaced (Whipple bumper), are used; new variations on both continue to be developed. The basic advantages of monolithic shields are their simplicity and low volume. Whipple bumper shields, however, will generally provide far better protection against high-velocity orbital debris than the same mass of monolithic shielding.

Monolithic shields are typically used to protect against small mass and lower-velocity impacts. In such impacts, the projectile's impact energy is low enough that it typically does not break up, and the shield is effective because its mass is sufficient to absorb and distribute the impact energy. At higher collision velocities, however, impacting objects often break apart on impact; at typical LEO collision velocities, an impacting object will generally melt or vaporize. Fragmented or melted impactors will either cause a large spherical crater or perforate the shield, depending on shield thickness. While monolithic shields can protect against high-velocity impacts, the monolithic shield thickness required to prevent perforation increases with approximately the two-thirds power of the collision velocity (see, for example, Swift, 1982; Cour-Palais, 1985; 1987).

At impact velocities greater than 2 to 3 km/s, a Whipple shield generally becomes more efficient (in terms of stopping debris per unit mass) than a monolithic shield. Experimental and theoretical evidence shows that at typical meteoroid impact velocities in LEO, Whipple bumpers provide protection equivalent to monolithic shields 10 to 20 times their mass (Swift, 1982). As illustrated in Figure 6-1, when a high-velocity projectile strikes a Whipple shield, the interaction with the bumper sends a shock wave through the projectile, initiating projectile breakup, melting, or vaporization. Smaller, reduced-velocity particles then travel between the bumper and the catcher and impact a larger area on the catcher. This spreads the total impact energy over a large area and ensures that each individual particle has relatively little energy or momentum, allowing the catcher to be much thinner than a monolithic shield.

Whipple bumper shields must protect against not only the high-speed particles that will break apart or vaporize on impact with the bumper but also the slower-moving objects that will simply perforate the bumper and strike the catcher still intact. A spaced shield with a thick monolithic catcher is thus required to protect against the entire range of debris ve-

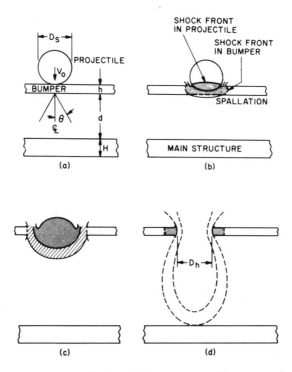

FIGURE 6-1 Projectile interacting with a spaced shield. (a) impact onto a thin bumper plate, (b) penetration, (c) subsequent formation of a spallation cone, and (d) loading transmitted by the cone to the catcher. SOURCE: Riney, 1970.

locities. In this type of shield, the thickness of the catcher is driven by the largest and fastest impactor expected to reach it without being broken apart by the bumper. The bumper is sized according to the highest-energy impactor expected, and the spacing between the two is designed to optimize distribution of projectile energy.

Several variations of Whipple bumpers are currently being developed and studied. These improvements, including the multilayer NEXTEL shield (Cour-Palais and Crews, 1990) and the mesh bumper shield (Christiansen and Kerr, 1993), reduce the mass of shielding needed to protect against a given environment and/or reduce the secondary ejecta produced by impacts into shields. Various shielding studies for the International Space Station, including a single aluminum Whipple bumper, a double aluminum bumper, and a stuffed Whipple bumper have also been conducted at ESA, NASA, and the National Space Development Agency of Japan (Christiansen, 1994; Ito, 1994; Lambert, 1994).

Active Protection

Active protection systems use sensors to warn of impending debris impact and mechanisms or motors to protect critical components or to move the spacecraft away from the potential impact. The only active protection schemes employed to date in space have involved using ground-based sensors to alert crewed or GEO spacecraft of potential collisions with cataloged objects; the spacecraft can then fire maneuvering rockets to safely avoid the objects. Other methods of active protection have been proposed, however. Some involve detecting oncoming small debris with on-board sensors and then either closing shutters over sensitive components or rotating the spacecraft so that the sensitive components are not struck. Still more technologically audacious active protection schemes involve shooting free-flying shields or directed energy weapons (lasers, plasmas, etc.) at oncoming debris to divert or fragment it before it strikes the spacecraft (Schall, 1993; Settecerri and Beraun, 1993).

All active protection mechanisms require advance detection and warning. Because debris may approach a spacecraft at velocities of greater than 10 km/s in LEO, most require warning when a potential impactor is hundreds of kilometers from the spacecraft to allow the spacecraft time to respond (i.e, safely maneuver, rotate within operating limits, fire at the impactor). The necessary detection and tracking capabilities to provide this warning can theoretically be supplied either by on-board sensors or by ground-based space surveillance systems.

Detecting debris with spacecraft-based remote sensors has been discussed in Chapter 2. Using such sensors even to detect debris is a difficult task; using them for collision warning is extremely demanding. In

---

BOX 6-2   Space-based Sensors and Collision Avoidance Maneuvers

Space-based sensors may not be capable of providing sufficiently timely and accurate warnings for most spacecraft to maneuver to avoid oncoming debris. For example, even if a 1,000-kg spacecraft in LEO were equipped with a sensor system capable of warning the spacecraft of an impending collision at 100-km distance with such accuracy that the spacecraft could avoid the debris by moving only 25 meters, the spacecraft would still have only about 5 seconds to move the 25 meters.

Such a maneuver would require a rocket engine with a thrust of 2 kN. (In comparison, typical station-keeping rockets have a thrust of 1 N.) If a 375,000-kg space station had to perform the same maneuver, it would need a rocket with 750 kN of thrust (about the same as the second stage of an Ariane 4 launch vehicle). Additionally, acceleration for such a maneuver would probably exceed the permissible G-loading on extended structures (such as solar panels). Increasing the distance at which the debris was detected would reduce both the propulsive thrust required and the resulting G-loading.

---

this role, the sensors must not only detect the oncoming object but also acquire enough position information to determine if it will hit the spacecraft, and they must do all of this fast enough to allow the spacecraft sufficient time to react (see Box 6-2). This problem is much simpler if the sensor is located on the spacecraft to be protected. However, the physical requirements of systems able to detect medium-sized debris at sufficient distances (and time) to allow action to be taken could be very demanding. For example, a space-based radar would require extremely high power levels; optics would have to be tens to hundreds of centimeters in diameter. The sensors would also require wide fields of view to detect all incoming debris.

Data from ground-based surveillance systems, on the other hand, have already been used for collision warning. As described in Box 6-3, space surveillance organizations use these data to project objects' future locations and to alert spacecraft if they will pass close by, or possibly collide with, another object. For this reason, ground-based collision warning systems, unlike space-based systems, have no problem providing sufficient warning time; they are, however, limited to warning of debris large enough to track from the Earth (currently the minimum size trackable is about 10 cm in diameter, as discussed in Chapter 2) and are limited in predictive accuracy.

Effective ground-based collision warning requires three main elements. First, the catalog must contain a significant fraction of the hazardous debris that could intersect the orbit of the spacecraft. Second, the system must provide sufficiently accurate information so that the ratio of false to real alarms does not require the spacecraft to make an excessive number of avoidance maneuvers. Finally, the spacecraft must be able to respond to the system's warning by moving out of the oncoming object's path (requiring both an effective maneuvering capability and a timely warning) or by employing other active protection measures.

Current collision warning capabilities do not meet these requirements. First, current catalogs are incomplete in size ranges less than 20 cm and thus cannot warn against the majority of potentially hazardous debris objects. In addition, uncertainty in predicting the future location of both objects potentially involved in a collision, due to the unpredictable nature of the upper atmosphere (described in Chapter 1), means that a high ratio of unnecessary to necessary maneuvers is inevitable. This uncertainty also prevents accurate prediction of collisions with sufficient advance warning for most current spacecraft to execute an avoidance maneuver.

Designers of such high-value missions as the International Space Station have explored the development of a collision warning system for debris down to 1 cm in diameter. Such a system would require a network of short-wavelength radars and propagation of the expected orbit of the objects with significantly more accuracy than currently achieved by either the SSS or the SSN. One estimate of the cost of such a network is $1 billion, with yearly operating costs of about $100 million (Loftus and Stansbery, 1993).

In summary, the sensor capabilities required for active protection

---

BOX 6-3 U.S. Space Shuttle Collision Avoidance

As a part of its reassessment of operating procedures after the *Challenger* accident, NASA developed a collision avoidance procedure for the Space Shuttle. Before the launch, the SSN analyzes the location of cataloged debris for the first four to five hours of the mission to determine if any will pass close to the shuttle. When the shuttle is in orbit, the SSN will notify NASA if a cataloged object is predicted to pass within 5 km radially or 25 km along the orbital track of the shuttle. If the predicted distance closes to 2 km radially or 5 km along the track, the shuttle will perform a collision-avoidance maneuver if it does not compromise either primary payload or mission objectives. The shuttle requires 45 minutes warning to plan and perform a collision-avoidance maneuver (General Accounting Office, 1990). From 1989, when this procedure was implemented, through February 1994, the shuttle received four notifications and performed three collision-avoidance maneuvers (Stich, 1994).

schemes are both technically challenging and expensive and thus would probably be used only for crewed or other highly valued spacecraft. Even with an effective collision warning system in place, however, many active protection methods may still be infeasible. The shoot-back schemes, for example, require far more power (perhaps tens of kilowatts) than most spacecraft can generate, and as discussed, maneuvering away from debris in the short-notice warning case can require high-thrust rockets and sturdy spacecraft design.

Operational Protection

Operational protection, including oversizing, redundancy, and mission/architecture design, is currently the most commonly used impact protection method in the spacecraft design community. Most operational protection schemes in place, however, were not implemented to deal with the debris hazard but rather to minimize the chance of mission failure due to component failure from any cause. To turn this logic around, a secondary advantage of operational techniques is that they not only protect against the debris hazard but also protect the spacecraft against failures unrelated to debris impact.

One operational protection approach is to design for "graceful degradation" so that, although a component may be operating out of its specified ranges, the degraded performance does not cause complete breakdown or mission failure. A typical example is thermal control, which depends on the reflective and absorptive properties of surface materials. Surface degradation can cause the temperature-controlled item to gradually approach and exceed designed temperature limits (usually the hotter limit). The operational protection approach is to design the thermal surface so that it initially provides more than enough thermal control and then to design the components underneath so that they too degrade gracefully when out of limits, thus increasing the safety margin. Oversizing can also be used for solar panels and other components to allow a given amount of degradation while retaining the required performance levels.

Another operational protection technique is redundancy, which is used primarily for electronic and propulsion components. This approach involves duplication of components in two or more places on the spacecraft so that if one component fails, another can take its place. Redundancy can even be applied to entire spacecraft constellations; the U.S. Global Positioning System utilizes this approach by maintaining more spacecraft in orbit than needed at any one time.

A third operational technique is to trade off system performance criteria with the orbital altitude and attitude in which the spacecraft will

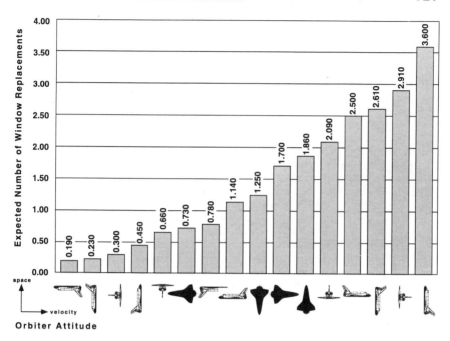

FIGURE 6-2 Expected number of window replacements for U.S. Space Shuttle at various orbital attitudes. SOURCE: NASA Johnson Space Center.

operate. One example of this is redesigning a mission so that a sensitive component functions facing the rear, rather than in the direction of motion, to reduce the flux on that component. Although mission requirements for the pointing of sensitive components make this approach infeasible for many missions, it can occasionally be employed. Figure 6-2 illustrates how orbital attitude changes the number of expected impacts on U.S. Space Shuttle windows. Space Shuttle Rule 2-77 states that the shuttle should use the orientation that causes the least number of window impacts unless it compromises mission objectives (NASA, 1993a). The space shuttle also uses orbits believed to have a lower flux of debris whenever possible; mission designs must keep shuttle orbital altitudes below 320 km whenever possible without compromising high-priority objectives (NASA, 1993b).

FINDINGS

Finding 1: Improved abilities to assess the hazard to spacecraft from orbital debris and, if necessary and feasible, to incorporate some protective measures to avoid spacecraft and component degradation or loss,

BOX 6-4 Designing for the Debris Environment: The RADARSAT Example

The Canadian RADARSAT is a 3,000-kg, three-axis stabilized spacecraft with about 5 square meters of frontal surface. It will be placed in a sun-synchronous orbit with an altitude of 790 km and an inclination of 98.5 degrees for a designed functional lifetime of five years. Because the orbital regime into which the spacecraft will be launched is believed to have a high debris flux and the initial RADARSAT design exposed many components to the space environment, the designers performed an assessment of the debris hazard to the spacecraft.

First, using NASA's EnviroNET, the designers predicted the approximate flux of space debris on the spacecraft, given the RADARSAT orbit parameters and configuration. This provided an estimate of the number of impacts expected on the leading face of the spacecraft with respect to debris size. The expected velocity and impact angle distributions for impacting particles were then determined using a NASA model of the LEO environment (Kessler et al., 1989). These indicated that one 1-mm particle and a larger number of smaller objects would impact the spacecraft's leading surface during its five-year mission. (Of course, as described in Chapter 3, any estimates of small debris populations in this orbital regime are based on significant extrapolations from existing data.) It was calculated that most impacts would occur with velocities in the 13- to 15.5-km/s range, with impact angles in the range of 5 to 30 degrees (measured from the direction of motion).

Using this information, spacecraft components were examined analytically to determine their vulnerability to impacts. The payload module contained electronic components behind honeycomb shear panels, so analysis of the module focused on determining whether the shear panels sufficiently shielded the electronics. Hypervelocity impact equations indicated that the components were adequately shielded. On the bus module, however, most of the sensitive equipment was mounted on the outside of the honeycomb shear panels and was therefore protected only by multilayer insulation (MLI) thermal blankets, separated from the shear panels by 15 to 25 cm. An analysis was performed to determine the vulnerability of each component (including not only electronic equipment, but also cable harnesses between boxes and propulsion subsystem hardware) and then to combine these numbers to determine an overall bus module vulnerability. The overall survivability against the meteoroid and orbital debris environment was calculated to be only 50% over the five-year functional lifetime.

To complement the computational analyses, some spacecraft components were subjected to impacts at the NASA's Hypervelocity Impact Research Laboratory. The tests were used to verify the assumptions made in the analyses and to determine the effectiveness of various shielding techniques. The test articles included different configurations of honeycomb shear panels, various thicknesses of electronic component walls, hydrazine lines, synthetic aperture radar waveguides, and wire bundles. MLI blankets were also tested, with and without reinforcement, to verify their shielding effectiveness. The impact angles were varied to simulate impacts on components on both the front and the sides of the spacecraft. Some results from the tests follow:

BOX 6-4—continued

- An MLI blanket 63.5 mm from a plate provided significant shielding against small projectiles. Adding Nextel or aluminum mesh to the blanket improved the shielding even further.
- The hypervelocity impact of a 1-mm aluminum projectile considerably damaged unprotected 24-gauge wire bundles, but negligibly damaged those protected by an MLI and Nextel shield.
- On a mass basis, Nextel and aluminum mesh performed similarly. However, because of the difficulty of adding three layers of aluminum mesh to the thermal blanket, one layer of Nextel was chosen as the baseline reinforcement.
- A 0.4-mm aluminum projectile impacting at 8 km/s could perforate hydrazine lines with stainless steel walls 0.51 mm thick.

A number of design changes increased the spacecraft's chances of survival in the predicted debris environment. These included adding a layer of Nextel to the MLI blankets of the bus module, thereby increasing the protection of the electronic boxes and the wiring harnesses mounted on the outside of the bus shear panels. Bus module components that were considered more vulnerable had their walls thickened. A gap between the bus module and the payload module was closed to protect a number of hydrazine lines. Shields were also added to some hydrazine lines to decrease the probability of direct hits. Finally, the forward cornerpost radiators of the bus were thickened and widened to shield some electronic components. Total spacecraft mass increased 17 kg from all of the shielding design improvements. These modifications, along with other changes from the evolution of the design, increased the predicted survivability of the spacecraft against micrometeoroids and orbital debris from 50 to 87 percent for its five-year mission.

are now available to spacecraft designers. Tools are becoming available to make these tasks easier, but it is important that spacecraft designers understand the assumptions that have been incorporated into them.

Finding 2: A spacecraft's basic structure should be the first line of defense against the debris hazard. If the spacecraft's structure does not provide sufficient protection, it may be necessary either to add additional shielding or to employ an active or operational protection scheme. Shielding can involve augmenting existing components or adding new shields. Selective local shielding of critical components can be a cost-effective means to reduce spacecraft vulnerability to debris or micrometeoroid impacts.

Finding 3: Active protection measures, such as movable shields and shutters, avoidance maneuvering, and direct attacks on incoming impactors, are costly and often technically challenging. They require advance detection and warning, which in turn may require improved sensor capabilities. Development of a collision warning system capable of protecting spacecraft effectively against all hazardous orbital debris would be expensive and challenging.

REFERENCES

Christiansen, E. L. 1993. Spacecraft Meteoroid/Debris Protection. Briefing presented to the National Research Council Committee on Space Debris Workshop, Irvine, California, November 18.

Christiansen, E.L. 1994. Shielding Status. Paper presented at the eleventh meeting of the Inter-Agency Space Debris Coordination Committee, Tsukuba, Japan, May 12–14.

Christiansen, E.L., and J.H. Kerr. 1993. Mesh double-bumper shield: A low-weight alternative for spacecraft meteoroid and orbital debris protection. International Journal of Impact Engineering 14(14):169–180.

Cour-Palais, B.G. 1985. Hypervelocity impact investigations and meteoroid shielding experience related to Apollo and Skylab. NASA CP-2360. Pp. 247–275 in Orbital Debris. Washington D.C.: National Aeronautics and Space Administration.

Cour-Palais, B.G. 1987. Hypervelocity impact in metals, glass and composites. International Journal of Impact Engineering 5:221–237.

Cour-Palais, B.G., and J.L. Crews. 1990. A multi-shock concept for spacecraft shielding. International Journal of Impact Engineering 10:499–508.

General Accounting Office (GAO). 1990. Space Program: Space Debris a Potential Threat to Space Station and Shuttle. Report to Congress, GAO/IMTEC-90-18. April. Washington, D.C.: GAO.

Ito, T. 1994. Hypervelocity Impact Test Results Recently Conducted by NASDA. Paper presented at the eleventh meeting of the Inter-Agency Space Debris Coordination Committee, Tsukuba, Japan, May 12–14.

Kessler, D.J., R.C. Reynolds, and P.D. Anz-Meador. 1989. Orbital Debris Environment for Spacecraft Designed to Operate in Low Earth Orbit. NASA Technical Memorandum 100471. April. Houston, Texas: National Aeronautics and Space Administration Johnson Space Center.

Kessler, D.J., R.C. Reynolds, and P.D. Anz-Meador. 1994. Space Station Program Natural Environment Definition for Design. NASA SSP 30425, Revision B. Houston, Texas: National Aeronautics and Space Administration Space Station Program Office.

Lambert, M. 1994. Space debris shield optimization. Paper presented at the eleventh meeting of the Inter-Agency Space Debris Coordination Committee, Tsukuba, Japan, May 12-14.

Lauriente, M., and W. Hoegy. 1990. EnviroNET: A Space Environment Data Resource. AIAA 90-1370. April. Greenbelt, Maryland: National Aeronautics and Space Administration Goddard Space Flight Center.

Loftus, J.P. Jr., and E.G. Stansbery. 1993. Protection of space assets by collision avoidance. Presented at the 44th Congress of the International Astronautical Federation, Graz, Austria, October 16–22. International Astronautical Federation.

NASA (National Aeronautics and Space Administration). 1993a. NASA Johnson Space Center Flight Rules. Flight Rule 2-77, Pp. 2-80a—2-80b. Houston, Texas: NASA Johnson Space Center.

NASA (National Aeronautics and Space Administration). 1993b. Space Shuttle Operational Flight Design Standard Ground Rules and Constraints. NSTS-21075, Rev. A, Level B,

Change 6, April 30, 4.2.4.2. Washington, D.C.: National Aeronautics and Space Administration.

Riney, T.D. 1970. Numerical evaluation of hypervelocity impact phenomena. Pp. 157–212 in High Velocity Impact Phenomena, R. Kinslow, ed. 1970. New York-London: Academic Press.

Schall, W.O. 1993. Active shielding and reduction of the number of small debris with high-power lasers. Pp. 465–470 in Proceedings of the First European Conference on Space Debris, Darmstadt, Germany, April 5–7 1993. Darmstadt: European Space Operations Center.

Sdunnus, H., and H. Klinkrad. 1993. An introduction to the ESA Reference Model for Space Debris and Meteoroids. Pp. 343–348 in Proceedings of the First European Conference on Space Debris, Darmstadt, Germany, 5–7 April 1993. Darmstadt: European Space Operations Center.

Settecerri, T.J., and J. Beraun. 1993. Laser debris sweeper for the Space Station Freedom. Pp. 471–477 in Proceedings of the First European Conference on Space Debris, Darmstadt, Germany, 5–7 April 1993. Darmstadt: European Space Operations Center.

Stich, J.S. 1994. Conjunction Summary for STS-26 through STS-61. NASA JSC Memo DM42/93-010. February 7. Houston Texas: National Aeronautics and Space Administration Johnson Space Center.

Swift, H.F. 1982. Hypervelocity impact mechanics. P. 215 in Impact Dynamics. New York: John Wiley & Sons.

7

Techniques to Reduce the Future Debris Hazard

There are many possible means of reducing the debris hazard to future space operations. These include actions taken as a spacecraft enters orbit (e.g., tethering rather than jettisoning lens caps and despin devices), during operations (e.g., reducing the amount of refuse ejected from crewed missions), and after its functional lifetime (e.g., depleting energy sources or moving the spacecraft into a disposal orbit). Some methods would cost very little, whereas others might be economically prohibitive for some missions. Their effectiveness also will vary, not only from method to method but also in how well a particular method will work in different orbital regions and with different space systems.

Methods to reduce the future growth in the debris population can be divided into two major categories: those that reduce only the short-term hazard and those that are also capable of reducing the long-term hazard. Measures that reduce the number of objects in orbit without reducing the total mass are effective only in diminishing the short-term hazard, because such measures do not reduce the total kinetic energy in orbit. It is this kinetic energy that constitutes the long-term collision hazard (Kessler and Loftus, 1994), so reductions in the long-term collision hazard require reducing the amount of mass in orbit. (This topic is discussed in greater detail in Chapter 8.)

There are two fundamental factors to consider when assessing methods to minimize the creation of new debris. The first is *how much the method will actually reduce the debris hazard to space operations*. The number of objects a particular method will prevent from being generated, the mass of those objects, and the threat those objects will pose to valuable orbital regions must all be considered. The second factor is *the difficulty and cost of implementing the debris reduction method.* This includes not only the development cost of any new hardware, but also the "opportunity cost" of any revenue lost or performance sacrificed in implementing the method. The choice of which methods to implement, when to implement them, and in what orbital regions they should be implemented typically involves a trade-off between these two factors.

MINIMIZING THE RELEASE OF MISSION-RELATED OBJECTS

As described in Chapter 1, there are three main types of mission-related debris: (1) objects released in spacecraft deployment and operations, (2) refuse from crewed missions, and (3) rocket exhaust products. Each of these debris types has very different orbital characteristics and size distributions. Together, they make up 13% of the total cataloged space object population; most of these objects are (as shown in Figures 3-4 and 3-5) located in orbital regions used by spacecraft. In addition, as discussed in Chapter 3, a large population of uncataloged mission-related debris also exists. Although ending the release of mission-related debris will obviously prevent the hazard from these objects from growing and further endangering future space operations, the balance between the costs and benefits of reduction actions varies greatly for the different types of mission-related debris.

Reducing the amount of *mission-related debris released in spacecraft deployment and operations* (e.g., clamps, covers for lenses or sensors, de-spin devices, pyrotechnic release hardware, wraparound cables) may be one of the easier ways of decreasing the future debris hazard to space operations. These objects make up the great majority of the cataloged mission-related debris population and typically have the longest orbital lifetimes of any mission-related debris. In the past, the practice has often been to simply jettison such items at separation from the launch vehicle or during appendage deployment. By using tethers or other simple devices, however, the release of most of these items can be avoided. Similarly, explosive bolts, which are commonly used to separate rocket upper stages, can be designed to not release large amounts of debris when activated. Because the parent spacecraft or rocket body would retain most objects, however, implementing such measures would not reduce the total mass of debris in orbit. (Chapter 8 discusses the significance of reducing mass in orbit.)

Measures to retain debris created during spacecraft deployment and operations are typically fairly easy to implement without affecting spacecraft operations. (Since the early 1980s, many such methods have been used on U.S. and other spacecraft.) The release of some types of mission-related debris during spacecraft deployment, however, may be more difficult to avoid. One example is dispensers for multiple spacecraft (e.g., the forward payload adapter on the Titan III and the SPELDA device used with the Ariane launch vehicle). Methods for retaining or deorbiting such items have not yet been developed, but development of such methods does not seem to be an inherently intractable problem.

Reducing the amount of *mission-related debris created during the course of crewed space activities* will have little effect on the overall debris hazard to space operations. Since human activities in space currently take place at low altitudes, the debris they release (mostly from intentional refuse dumping and extravehicular activities) experiences rapid orbital decay and does not contribute to the long-term debris population. Although there are a number of possible methods to further reduce the hazard to space operations from such debris (e.g., bringing the refuse back to Earth during scheduled crew rotations or attaching a drag augmentation device to speed its orbital decay), implementing such methods will not reduce the overall long-term debris hazard. However, since this debris contributes to the short-term hazard in an area containing valuable spacecraft, the use of low-cost methods of debris reduction (if such methods are available) appears to be worthwhile.

Curtailing the release of *exhaust products of solid rocket motors* will also do little to reduce the debris hazard to space operations. As discussed in Chapter 3, solid rocket firings produce a vast number of very small (<10–micron) debris, but their orbital lifetimes are fairly short due to the strong effect of perturbing forces such as solar radiation pressure; less than 5% will remain in orbit after a year. In addition (as described in Chapter 4), the surface degradation these particles can cause is not a major hazard to functional spacecraft.

The only methods of meaningfully reducing this population would be either to restrict solid rocket motor firings in orbit or to alter the composition of solid rocket motor fuel. Because either action would impose cost increases or performance reductions on many space activities, and the lifetime of these exhaust particles and the potential damage that they can cause to functional spacecraft are so small, it seems clear that neither step is yet warranted at present. It is not yet clear, though, whether anything should be done to limit the population of 1-cm and larger pieces of slag (discussed in Chapter 1) that are also believed to be ejected during solid rocket burns. Whereas the larger size and longer orbital lifetimes of these particles may make them a greater hazard to spacecraft than the small aluminum oxide particles, too little is currently known about them (in particular, how many are typically produced in a solid rocket motor firing) to determine if there is any need to search for ways to prevent their creation.

SAFEGUARDING THE PHYSICAL INTEGRITY OF ROCKET BODIES AND SPACECRAFT

Reducing the Creation of Debris from Explosions

Fragmentation debris makes up 42% of the cataloged space object population and probably a much larger fraction of the uncataloged population. Since there have been only two confirmed space object breakups to date due to collisions (both intentional military tests), the vast majority of this debris is believed to have been created in explosive breakups of spacecraft and rocket bodies. This population of debris spans all size ranges and is distributed widely, although concentrated near the orbits in which it was created. Figure 7-1 projects how a typical explosion in LEO (producing 300 cataloged objects) could moderately increase the spatial density of cataloged objects in orbits hundreds of kilometers above

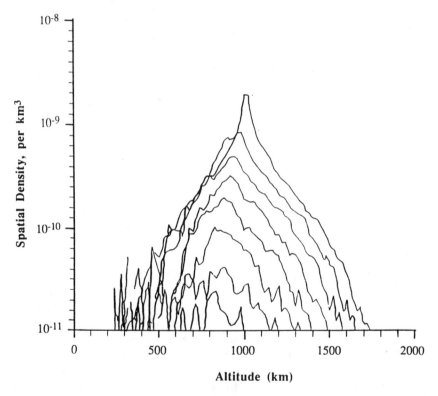

FIGURE 7-1   Predicted effect of satellite breakup at 1000 km. Top curve is initial spatial density distribution with altitude. Time interval between lower curves is 400 years. SOURCE: Kessler, 1991.

and below the breakup altitude. This increase in spatial density can persist for long periods of time; the higher the altitude, the longer will the spatial density remain elevated. Since explosions can produce a considerable amount of large and medium-sized debris with potentially long orbital lifetimes, reducing the creation of debris from explosions will clearly have a major effect in containing future growth in the debris hazard. A reduction in the frequency of explosions can be achieved by passivating spacecraft and rocket bodies.

Passivation of Spacecraft

Debris from spacecraft explosions makes up about 12.5 percent of the cataloged space object population. Spacecraft can explode both during and after their functional lifetime for a wide variety of reasons, including propellant tank explosions, thruster malfunctions, tank failures due to the impact of small debris, battery ruptures, accidentally induced high rotation rates, other degradations of the structure, or deliberate explosions. There are correspondingly many possible measures to prevent spacecraft breakups. There is no one single remedy, and there is probably no possible way to avoid *all* future spacecraft breakups: despite safeguards, a residual number of spacecraft breakups will continue to generate debris, if at a reduced level.

However, spacecraft designers can take a general system-level approach to prevent accidental spacecraft breakups. The approach is (1) to determine all potential sources of stored energy remaining on a spacecraft late in its active life; (2) for each source, to provide a method of dissipating the stored energy in a benign manner; and (3) to activate these means at the end of the spacecraft's functional lifetime (i.e., "passivate" the spacecraft). Protecting the spacecraft from debris impact damage, as well as other methods to increase spacecraft survivability, can help ensure that the spacecraft is capable of carrying out passivation measures at its EOL.

The "passivation" approach described above can be applied to numerous spacecraft subsystems. For example, spacecraft batteries are sources of stored energy believed responsible for a few breakups. To prevent such breakups, designers can implement a battery management system that ensures that the batteries will be left in a completely discharged state at the end of the spacecraft's functional lifetime and will be short-circuited to prevent recharging. Implementation of this system would prevent inadvertent overcharging, which can lead to battery rupture and potentially break up spacecraft. Another example of this approach would be to ensure that all residual propellants and stored pressurized gas in the spacecraft are vented at the end of the spacecraft's functional lifetime—if possible, in a manner that moves the spacecraft into an orbit that reduces its long-term contribution to the debris hazard.

Ending or reducing deliberate spacecraft breakups would also, of course, reduce the spacecraft fragment population. Historically, spacecraft have been broken up deliberately for structural testing, to destroy sensitive equipment so that it would not be recovered by hostile forces, and in antisatellite weapons tests (Johnson and McKnight, 1991). Deliberate breakups are believed to account for slightly more than one-third of all spacecraft breakups. Another 20 percent of all spacecraft breakups may be due to the *unintentional* detonation of on-board self-destruction systems. Combined, these types of breakups are the source of approximately 6 percent of the current cataloged space object population. Deliberate breakups of spacecraft about to reenter the atmosphere do not contribute greatly to the debris hazard; the debris created in such events is typically ejected into orbits that decay rapidly. Fragments from intentional breakups at high altitudes (>600 km) can, however, remain in orbit for thousands of years or more. Ensuring that any future deliberate spacecraft breakups are not carried out in high orbits would help contain the future debris hazard.

Passivation of Rocket Bodies

Debris generated through the explosive breakup of liquid-fueled rocket bodies after they have completed their missions makes up 25 percent of the cataloged space object population, and probably a large fraction of the uncataloged large and medium-sized debris population. Rocket body breakups are believed to be caused most often by the residual propellant (as much as several hundred liters) that may remain in the rocket body's fuel and oxidizer tanks at the end of a mission. Explosions that break up rocket bodies are caused most often by accidental mixing of the components of this residual propellant or by physical factors such as overpressure.

Accidental mixing occurs most commonly in rocket bodies that store fuel and oxidizer in thin tanks with a common bulkhead. During ground handling and launch, a positive pressure difference exists between the oxidizer tank and the fuel tank, but after spacecraft separation, this pressure difference can vanish due to leaks in pipes and valves, resulting in damage to the bulkhead. Fuel and oxidizer are then able to mix through the damaged bulkhead, leading to an explosion. The bulkhead also can rupture from corrosion or thermal stress; thermal stressing of a fuel tank bulkhead may have led to the breakup of seven Delta rocket bodies. Fragmentations caused by accidental fuel mixing can be extremely energetic, because of the large amount of fuel that may be involved.

Rocket body explosions also can be generated by nonchemical means, such as overpressure leading to propellant tank rupture. Overpressure may occur for a number of reasons, including propellant heating and failure of pressure relief valves. Explosions caused by nonchemical means are often less energetic than those caused by propellant mixing. Since explosions caused by overpressure cause no transient stresses, theoretically the propellant tank will tear along lines of weakness, generating few, if any, fragments, and the additional velocity imparted to any fragments should also be low (Fucke, 1993). However, the 1986 explosion of an Ariane third stage, which is believed to have been caused by over-pressurization, produced a record number of cataloged fragments, and explosions generated by nonchemical means probably caused seven of the ten largest fragmentation events recorded (all with more than 225 cataloged fragments).

Launch vehicle builders have developed a number of methods to reduce a rocket body's potential for explosion. In general, the methods involve either (1) depletion burns after the rocket body separates from the spacecraft or (2) venting of residual propellant. Although these passivation measures will not eliminate propulsion-related breakup events (i.e., breakups that occur during rocket ignition and propulsion), such events are rare for orbital rocket bodies.

In *depletion burns*, the engine is reignited after completion of the staging process and operated under normal conditions until its propellant is depleted. In principle, depletion burns can shorten the rocket body's orbital lifetime, although past burns of some rocket bodies have increased orbital lifetime. (See the discussion of orbital lifetime reduction later in this chapter.) Such a maneuver typically requires using the rocket body's battery for power and its auxiliary thrusters for attitude control. To gain the maximum lifetime reduction from such a maneuver, the depletion burn should be carried out near the orbit's apogee; to prevent the con-tamination of nearby spacecraft, some rocket bodies may have to retain the capability to make such burns for several hours after staging. Cur-rently, some rocket bodies are capable of performing depletion burns for a short time after spacecraft delivery, and most other rocket bodies would require only minor modifications to be able to perform depletion burns.

Venting of residual propellant can be achieved either by blowing the propellant out through valves or by evaporating and venting it. To vent residual propellant, a rocket body generally requires pressure relief valves (usually activated by firing pyrotechnic devices) and venting pipes. The advantage of venting is that it does not require reignition or auxiliary thrusters. The Ariane rocket bodies (see Box 7-1) now vent their residual propellant.

Residual propellant from the main rocket engines is not the only

---

BOX 7-1   Ariane Passivation

The Ariane 1 through Ariane 4 are three-stage launch vehicles with cryogenic third stages. The third-stage liquid-oxygen and liquid-hydrogen tanks use a common bulkhead. On average, 120 kg of liquid hydrogen and 160 kg of liquid oxygen remain after third-stage engine cutoff.

Passivation measures for the third stage ensure full depletion of the residual propellant. Venting of the tanks begins when pyrotechnic devices fire to activate the pressure relief valves and venting pipes that were installed for this procedure. Depletion is timed so that the pressure difference between the two tanks meets the bulkhead design requirement throughout the procedure.

---

cause of rocket body breakup. In several cases, debris has been generated by the explosion of residual propellant for the auxiliary engines used to maintain three-axis control during transfer orbit segments and to provide axial acceleration prior to rocket body reignition. Propellant venting and depletion burns also can be used to avert such explosions. Finally, batteries and other pressurants on rocket bodies are sources of energy that can lead to breakups. These can be passivated in the same manner as they would be on spacecraft.

Reducing the Creation of Debris from Degradation

The products of spacecraft surface deterioration include paint flecks and other surface materials that come loose from a space object under the influence of the space environment. Very few of these items are large enough to be cataloged; the vast majority are small. The few cataloged objects believed to be released due to surface degradation have had high ratios of cross-sectional area to mass and have experienced relatively rapid orbital decay. The vast numbers of small particles released due to surface degradation are also suspected to have high ratios of cross-sectional area to mass and thus fairly short orbital lifetimes (as discussed in Chapter 3). However, since a typical paint fleck may have a mass of only 10^{-6} gram, the deterioration of even minor amounts of surface material can rapidly replenish the orbiting population. As discussed in Chapter 4, these particles can cause surface degradation and can also potentially damage unprotected spacecraft components such as optics, windows, and tethers.

Much has been learned from LDEF and other experiments about the effect of the space environment on various substances; thermal coatings and treatments that reduce surface charge buildup and have other long-life properties are now generally available. Although spacecraft designers commonly avoid using paint or other surface materials that significantly deteriorate during the spacecraft's functional lifetime, they generally do not require that the surface coatings survive intact long after the spacecraft's functional lifetime. The situation is similar for rocket bodies, although in this case the surface materials may be required to remain intact in space only for a few hours or days (although they must survive in the often harsh environment of the launch pad for long periods of time). Education of spacecraft and rocket body designers about the hazards caused by surface degradation and the preventive measures available may be an inexpensive means of reducing the creation of this type of debris.

REDUCING THE CREATION OF DEBRIS FROM COLLISIONS

Two main approaches could theoretically be employed to reduce the long-term creation of debris from collisions. These are (1) to decrease the number of collisions by employing collision avoidance techniques or (2) to remove objects capable of causing collisions away from crowded orbital regions. (Limiting the *number* of objects in orbit without reducing mass is not sufficient to reduce the long-term potential for collisions, because such reductions do not affect the total kinetic energy in orbit available to cause collisions.) The problem with the first approach is that, as discussed in Chapter 6, current collision warning systems are ineffective, and the development of effective systems would be both technically challenging and costly. Even if an effective collision warning system were implemented, it would probably not be of use in preventing breakups of either nonfunctional spacecraft or other debris (because such objects are incapable of maneuvering to avoid a collision). Consequently, removing debris from crowded orbits may be the only practical alternative.

There are four techniques that can move debris from heavily trafficked orbits: (1) deorbiting (the deliberate, forced reentry of a space object into the Earth's atmosphere by application of a retarding force, usually via a propulsion system) at EOL; (2) orbital lifetime reduction (accelerating the natural decay of spacecraft and other space objects to reduce the time that they remain in orbit) at EOL; (3) moving objects into less populated "disposal" orbits at the end of their functional lifetime; and (4) active removal of debris from orbit.

Deorbiting/Lifetime Reduction

Although breakup fragments outnumber all other types of orbital debris, rocket bodies and (to a lesser degree) spacecraft have by far the largest fraction of mass and cross-sectional area in orbit. Most collisions, therefore, will involve these objects. The abandonment of rocket bodies and spacecraft in Earth orbits—especially in long-lifetime orbits such as GEO, circular orbits higher than 800 km, some GTOs, and some Molniya-type orbits—greatly increases the long-term potential for future collision. Possible techniques for deorbiting or accelerating the decay of these objects include the use of retrograde propulsion, natural perturbing forces, and drag augmentation devices.

Retrograde propulsion burns can be performed with dedicated small rocket thrusters or, as previously discussed, through a depletion burn of excess on-board fuel with existing rockets. A retrograde burn can be used either (1) to achieve a controlled deorbit, in which the rocket body or spacecraft is directed to impact (or burn up during reentry) at a predetermined location over an ocean or another uninhabited area, or (2) to maneuver the rocket body or spacecraft to an orbit with a lower perigee, which will lead to a shorter orbital lifetime followed by uncontrolled atmospheric reentry and burnup.

Both spacecraft and rocket bodies may require some modifications to carry out deorbiting or lifetime reduction maneuvers. Some spacecraft may not have attitude or orbit control systems capable of performing EOL burns; such systems are necessary to execute reorbiting or lifetime reduction maneuvers. Rocket bodies may need enhanced batteries, attitude control, and command systems in order to remain functional long enough to perform the retarding thrust maneuver. (This is particularly important for rocket bodies in orbits near the spacecraft they have just released into orbit; such rockets must often perform the retarding thrust

---

BOX 7-2 Examples of Lifetime Reduction Maneuvers

Although the Soviet Union deorbited many of its space stations and other large spacecraft, these maneuvers were not performed to reduce the hazard to other spacecraft but rather to avoid creating a hazard to people and property on the ground. Spacecraft in planned LEO constellations, such as those being developed by the Iridium and Teledesic organizations, may become the first spacecraft to carry out lifetime reduction maneuvers specifically to reduce the debris hazard. Current Iridium plans are to use retrograde propulsion burns to accelerate the orbital decay of the spacecraft from their initial 780-km orbits.

maneuver several hours after separation from the spacecraft.) Both rocket bodies and spacecraft would require sufficient fuel to perform these maneuvers.

Figure 7-2 shows the change in velocity (ΔV) and propellant mass fraction required to perform a deorbiting or lifetime reduction maneuver from various circular low Earth orbits. (The propellant mass fraction is the mass of the propellant divided by the total mass of the space vehicle, including propellant.) As can be seen, the mass of fuel required to deorbit a spacecraft or rocket body is greater than the amount needed to reduce its orbital lifetime. A rocket body with a ratio of cross-sectional area to mass of 0.01 m^2/kg in an 800-km circular orbit, for example, would require about half the amount of propellant to reduce its orbital lifetime to 10 years than it would require to deorbit. Performing either maneuver would remove a large, long-lived (up to hundreds of years) hazard from LEO, but the extra fuel required for either maneuver would directly reduce the launch vehicle's or spacecraft's payload capacity, making it less capable and putting it at a disadvantage with competitors that are not carrying out such maneuvers.

Natural perturbing forces can sometimes be used to reduce rocket body orbital lifetimes. Atmospheric drag is obviously a perturbing force with a major effect on the orbital lifetimes of objects that pass through low-altitude regions. Figure 1-6 illustrates how initial altitude can affect the orbital lifetime of various space objects in circular orbits. Figure 7-3 illustrates how orbital lifetimes for objects in elliptical orbits can vary even more sharply, depending on their initial perigee altitude. Clearly, rocket bodies launched into transfer orbits with low perigees experience much more rapid orbital decay than those launched into orbits with higher perigees; when possible, this can be a very effective means of limiting the orbital lifetimes of rocket bodies in highly elliptical orbits.

More subtle gravitational perturbations can also affect the orbital lifetime of objects in geostationary transfer orbits with perigees below about 300 km. Careful selection of the orbit's orientation with respect to the Sun and Moon (by launching at a particular time of day) can cause lunar-solar perturbations to lower the orbit's perigee. Figure 7-4 shows how the orbital lifetime of a rocket body varies depending on the initial sun angle. This technique could be a low-cost option to accelerate orbital decay from certain missions, but it can require major design changes for other missions; a comprehensive analysis is needed for each particular mission to examine possible conflicts with other requirements.

Finally, *drag augmentation devices* can be used to accelerate the orbital decay of rocket bodies or spacecraft. Drag augmentation, which would be effective only in low-altitude orbits, would involve deploying a device to increase the surface area, and thus the drag, of a space object. Figures

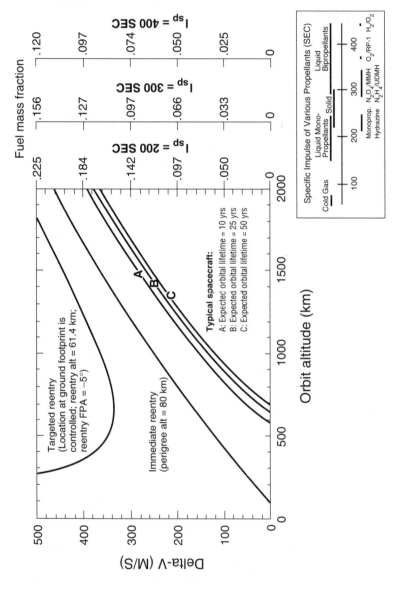

FIGURE 7-2 Delta-V (ΔV) requirements for disposal from circular orbit. Ratio of cross-sectional area to mass = 0.01 m²/kg. Solar activity index = 130 SFU. SOURCE: NASA/Reynolds.

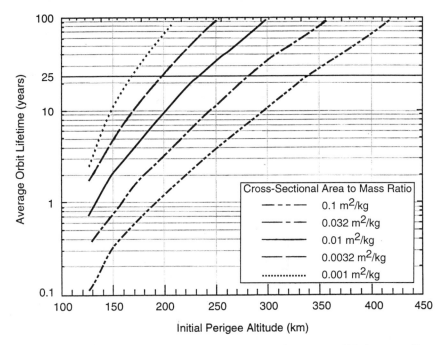

FIGURE 7-3 Average orbital lifetime for GTO, inclination = 27.5 degrees. Random choice for initial argument of perigee and right ascension of ascending node of orbit plane. SOURCE: National Aeronautics and Space Administration.

1-6 and 7-3 showed how increasing an object's ratio of cross-sectional area to mass can greatly reduce its orbital lifetime. Most drag augmentation concepts involve inflatable balloons, which are fairly simple to deploy and can produce a large surface area without a great mass penalty. One problem with balloon devices is that they will rapidly be punctured by small debris; this problem might be solved, however, if a proposed method involving post inflation solidification could be implemented. Alternatively, a non-balloon drag augmentation device, which might be more complex to deploy, could be used. Perhaps a greater problem with these methods is that while they reduce an object's orbital lifetime, they also increase the object's cross-sectional area; the effect may be that the total exposure to collisions is not significantly reduced.

Disposal Orbits

Deorbiting or meaningfully accelerating the orbital decay of spacecraft or rocket bodies from most widely used high-altitude orbital re-

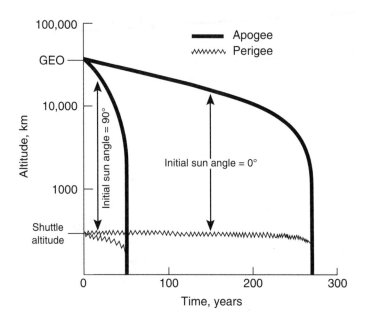

FIGURE 7-4 Shuttle-launched LEO to GEO transfer stage lifetimes. Initial sun angle is the angle between the transfer stage's original orbit and the sun. SOURCE: Loftus et al., 1992.

gions would be prohibitively costly. One means of removing objects from these regions is to reorbit them into "disposal orbits" at the end of their functional lifetime. This leaves the objects in Earth orbit but removes them from regions where they would pose a direct collision hazard to functional spacecraft. Disposal orbits must typically be far enough from the initial orbit that orbital perturbations do not take the reorbited objects back through their initial orbit, although stable disposal orbits within widely used orbital regions have also been proposed. Reorbiting into a disposal orbit typically requires two propulsive burns at the end of a spacecraft's or rocket body's functional lifetime.

Disposal orbits have been proposed for GEO and other orbital regions, including higher LEO orbits and semisynchronous orbits. A considerable number of GEO spacecraft and some spacecraft in semisynchronous orbits have already performed reorbiting maneuvers to reduce the future debris hazard in those orbits. Spacecraft in the semisynchronous Global Positioning System constellation have performed end-of-life reorbiting maneuvers to disposal orbits from approximately 220 to 810 km above or 95 to 250 km below their initial orbits. In GEO, spacecraft owned by many nations and organizations have carried out reorbiting maneuvers; Figure 7-5 shows the approximate number of GEO reorbiting maneuvers compared to the number of spacecraft launched to GEO by year. These reorbiting maneuvers have typically placed spacecraft in orbits from 50 to 1,000 km above GEO, though a few spacecraft have been reorbited to orbits below GEO.

Moving a space object into a disposal orbit reduces the collision hazard in the object's initial orbital region, but increases the collision hazard in its new orbital region. Objects moved to disposal orbits can still contribute to the debris hazard in their original orbit, however, since debris generated through collisions or explosions that take place in disposal orbits may intersect the original orbit. (Increased implementation of passivation measures should, however, result in fewer explosions in future disposal orbits.) This is particularly important in high-altitude regions, where an explosion or collision can send a large number of fragments far above and below their initial orbital altitude. Figure 7-6 shows a prediction of the effect of a spacecraft breakup in GEO on the large object flux in nearby altitudes. (This figure should not be compared with the GEO flux depicted in Figure 4-5 because that figure depicts only the flux due to cataloged [typically larger than 1 m in diameter] objects.) Clearly, the

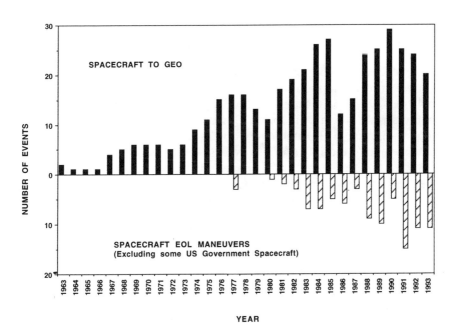

FIGURE 7-5 GEO spacecraft reorbiting maneuvers. SOURCE: Kaman Sciences Corporation.

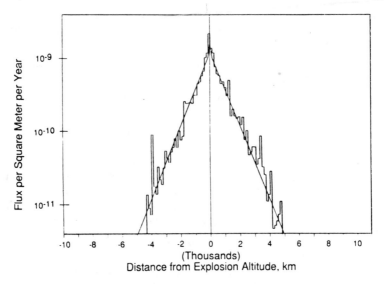

FIGURE 7-6 Calculated flux of 500 debris fragments >10 cm from a satellite breakup near GEO altitude. SOURCE: Friesen et al., 1992.

farther the disposal orbit is from the original orbit, the smaller will be the amount of debris generated in the disposal orbit that intersects the original orbit.

The "cost" of reorbiting to a disposal orbit is usually measured by the amount of fuel required to perform the maneuver. For rocket bodies, this fuel translates into reduced payload capacity; for spacecraft, it means that less mass is available for either payload or (more typically) for station-keeping fuel. The fuel required to move to a disposal orbit a certain distance above the initial orbit decreases with increasing initial orbital altitude. Figure 7-7 shows the change in velocity required to reach disposal orbits from three orbital regions. (This figure also explains why fragments from explosions at high altitudes become more widely distributed than fragments from explosions at low altitudes: given the same ejection velocity, the fragments at high altitudes will travel farther.) Of course, in addition to fuel, the spacecraft or rocket body must have the necessary propulsion and attitude-control capabilities to perform this maneuver.

For each disposal orbit, the reduction of the debris hazard to functional spacecraft should be balanced against the cost of moving objects to the disposal orbit at the end of their functional lifetimes. In addition, use of a disposal orbit should also be weighed against other possible debris reduction methods that may remove the object entirely from orbit. Within

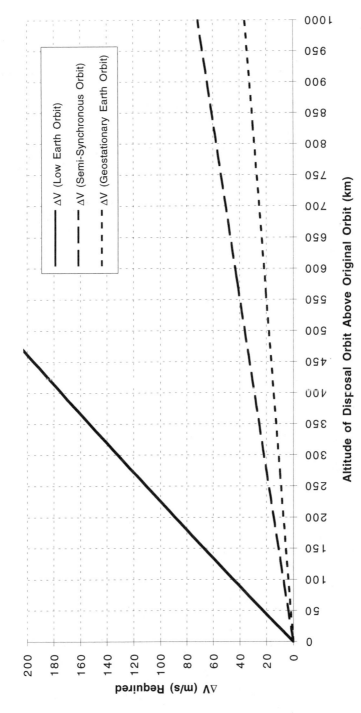

FIGURE 7-7 Delta-V (ΔV) costs for disposal orbits.

LEO, where it is often feasible to deorbit or accelerate the orbital decay of spacecraft or rocket bodies at EOL and where there is a lack of sufficiently unpopulated regions, disposal orbits do not seem to be a viable option. In GEO, however, where deorbiting and orbital lifetime reduction are infeasible, disposal orbits may be a viable option.

A number of potential GEO disposal orbits that would not require significant EOL maneuvering have been proposed. One concept is to group non-functional spacecraft at the "stable points" of the geostationary ring (above 75°E and 105°W longitude). Theoretically, objects at these locations will not drift along the geostationary ring and thus will not endanger spacecraft elsewhere in GEO. This disposal scheme, however, renders the stable points unusable for functional spacecraft and may not effectively reduce the overall debris hazard in GEO. In addition, the stable points are only mildly stable; a velocity change of only about 1.5 m/s is enough to send objects at the stable points moving along the geostationary arc. In addition, spacecraft at the stable points still experience the 15-degree GEO inclination variation cycle (as described in Chapter 3), and may develop high velocities relative to other spacecraft at the stable points.

Another idea has been to launch spacecraft to the 7.3-degree-inclination GEO "stable plane." The major perturbing forces on spacecraft in this orbit cancel each other, so spacecraft orbiting in the stable plane tend to remain in that plane (Friesen et al., 1993) and collision velocities between uncontrolled spacecraft in the plane would be only a few meters per second. In addition, spacecraft in the stable plane would not require station-keeping propellant to prevent north-south oscillations (which normally account for 95% of a geostationary spacecraft's propellant expenditure). However, objects in the stable plane move at velocities of close to 400 m/s relative to objects in geostationary orbit. Use of the stable plane would thus significantly reduce the debris hazard only if most GEO objects were in the stable plane. However, since spacecraft in the stable plane would not be *geostationary* and thus would not have the advantages of remaining above a particular point on the Earth, it seems unlikely that the majority of spacecraft operators would move their spacecraft to the stable plane. Even if all new GEO spacecraft were launched to the stable plane, they would still face a collision hazard from the objects that currently exist at other inclinations at the GEO altitude, although the overall collision hazard would be lower than if current practices were continued.

Since major problems exist in schemes to reorbit within the GEO altitude, the reorbiting of GEO spacecraft into disposal orbits with altitudes above or below GEO is the only practical method of removing mass from GEO. Although orbits above GEO may eventually decay into

GEO, the time frame for such a decay is believed to typically be on the order of tens of thousands of years or more. Objects reorbited below GEO, however, would pose an immediate (if low-level) hazard to objects in transfer orbits to GEO. Analysis of the long-term stability of orbits beyond GEO is still under way, but preliminary analysis shows that the use of orbits 300 km above GEO appears to be a minimum for effectively reducing the debris hazard (Chobotov, 1990; Yoshikawa, 1992). The 300-km figure is a minimum to ensure that (1) uncontrolled spacecraft do not physically interfere with controlled spacecraft in GEO and (2) functional spacecraft can change their operational longitude without interference. The use of disposal orbits a minimum of 300 km above GEO was recently recommended by the Ad Hoc Expert Group of the International Academy of Astronautics (International Academy of Astronautics, 1992).

There are disagreements even among experts about the value of using GEO disposal orbits. It is clear that (1) use of a disposal orbit will reduce the amount of mass in GEO and will thus reduce the GEO collision probability; (2) the hazard from objects that are not removed from GEO will persist for millennia; and (3) transfer to a disposal orbit above GEO is a simple maneuver requiring only as much propellant as is typically required by spacecraft for three months' station keeping. However, it is also true that (1) removing objects a few hundred kilometers above GEO only moves the hazard to a slightly wider band; it does not completely and permanently eliminate the hazard the object poses to spacecraft in the geostationary band; and (2) the hazard from simply leaving a spacecraft or rocket body in the geostationary orbit appears to be extremely low at the present time.

Active In-Orbit Debris Removal

The active removal of large debris (such as nonfunctional spacecraft and rocket bodies) from orbit has often been proposed as a means of reducing the debris hazard. The removal of large objects would require some kind of space vehicle dedicated to this purpose; all indications are that the cost of such a vehicle would be prohibitive, especially when the small reduction in the debris hazard that it could achieve is considered. (One study predicted a best-case cost of more than $15 million for each piece of debris in LEO removed, not counting the cost of developing an orbital maneuvering vehicle [Petro and Ashley, 1989].) Even ingenious schemes involving the use of tethers to deorbit large objects would likely be very costly.

A number of on-orbit active removal schemes for small debris also have been proposed, including the removal of small debris with "debris sweepers" (large foam balls or braking foils that impact with smaller debris) and the ground- or space-based laser evaporation of debris surface material to deorbit small debris. The sweeper scheme seems technically difficult, demonstrably inefficient, hazardous to functional spacecraft, and risks producing more small objects than it eliminates. The laser concept, although interesting, requires costly new technology, and its feasibility has not yet been proven. In general, there is currently no technology able to remove small debris efficiently, and any foreseeable schemes look very costly.

FINDINGS

Finding 1: The future debris hazard can be significantly ameliorated without exorbitant costs by ending or sharply reducing the number of breakups of spacecraft and rocket bodies and, to a lesser extent, by reducing the amount of mission-related debris released in spacecraft deployment and operations. Methods to achieve both these goals exist, are relatively inexpensive, and have been proven in orbit. While implementing these methods will reduce the total *number* of objects in orbit, it will not, however, significantly reduce the total *mass* of objects in orbit.

Finding 2: Deorbiting or accelerating the orbital decay of spacecraft and rocket bodies at the end of their functional lifetimes can reduce the total amount of mass and cross-sectional area in orbit. The difficulty and cost of such maneuvers vary depending on the initial orbit, the capabilities of the space vehicle involved, and the desired reduction in orbital lifetime. In general, significant reductions in orbital lifetime can be achieved with much less fuel than deorbiting would require.

Finding 3: Reorbiting spacecraft and rocket bodies into disposal orbits can reduce the debris hazard in their original orbit, but it is not a permanent solution since the debris remains in Earth orbit. Decisions to use a disposal orbit must balance the reduction in the long-term hazard to functional spacecraft against the cost of the maneuver, including the cost of carrying the required fuel and/or the need for premature shutdown. Disposal orbits are not a useful alternative within LEO; opinion within both the committee and the space debris community is divided as to whether they should be used by all spacecraft and rocket bodies in GEO.

Finding 4: The active removal of debris will not be an economical means of reducing the debris hazard in the foreseeable future. Design of future spacecraft and launch vehicles for autonomous deorbiting, lifetime reduction, or reorbiting is a far more economical means of reducing the collision hazard.

REFERENCES

Chobotov, V.A. 1990. Disposal of spacecraft at end-of-life in geosynchronous orbit. Journal of Spacecraft and Rockets 27(4):433–437.

Friesen, L.J., A.A. Jackson IV, H.A. Zook, and D.J. Kessler. 1992. Results in orbital evolution of objects in the geosynchronous region. Journal of Guidance, Control, and Dynamics 15(1):263–287.

Friesen, L.J., D.J. Kessler, and H.A. Zook. 1993. Reduced debris hazard resulting from a stable inclined geosynchronous orbit. Advances in Space Research 13(8):231–241.

Fucke, W. 1993. Fragmentation experiments for the evaluation of the small size debris population. Pp. 275–280 in Proceedings of the First European Conference on Space Debris, Darmstadt, Germany, 5–7 April 1993. Darmstadt: European Space Operations Center.

International Academy of Astronautics Committee on Safety, Rescue, and Quality. 1992. Position Paper on Orbital Debris. August 27. Paris: International Academy of Astronautics.

Johnson, N.L., and D.S. McKnight. 1991. Artificial Space Debris. Malabar, Florida: Krieger Publishing Company.

Kessler, D.J. 1991. Collision cascading: The limits of population growth in low Earth orbit. Advances in Space Research 11(12):63–66.

Kessler, D.J., and J.P. Loftus, Jr. 1994. Orbital debris as an energy management problem. Paper presented at the 31st Plenary Meeting of the Committee on Space Research (COSPAR), Hamburg, Germany, July 14. To be published in Advances in Space Research.

Loftus, Joseph P., Jr., D.S. Kessler, and P.D. Anz-Meador. 1992. Management of the orbital environment. Acta Astronautica 26(7):477–486.

Petro, A., and H. Ashley. 1989. Cost estimates for removal of orbital debris. Pp. 183–186 in Progress in Astronautics and Aeronautics: Orbital Debris from Upper-Stage Breakup, Vol. 121, J.P. Loftus, ed. Washington D.C.: American Institute of Aeronautics and Astronautics.

Yoshikawa, M. 1992. Long-term analysis for the orbital changes of debris. Pp. 2403-2408 in 18th International Symposium on Space Technology and Space Science (ISTS), Kagoshima, Japan, May 17-23. Tokyo: ISTS.

8

The Future Orbital Population and the Effectiveness Of Debris Reduction Measures

THE FUTURE ORBITAL POPULATION

As described in Chapter 1, the number of cataloged space objects has increased nearly linearly since 1960, at an average rate of about 220 objects per year. New spacecraft, the mission-related debris and rocket bodies associated with those spacecraft, and the fragments caused by the breakup of objects in space have contributed to this growth. Without the effect of orbital decay caused by atmospheric drag, the increase in the space object population would have been much larger. To date, more than 15,000 cataloged objects (about twice the current cataloged population in orbit) have decayed into the atmosphere and, at the peaks of the 11-year solar activity cycle, the overall losses of cataloged space objects have occasionally outnumbered the increases, resulting in an overall decline in the cataloged population.

It is this balance between the creation of new debris and the orbital decay of existing debris that will determine the magnitude and distribution of the future debris population. For each altitude region, this balance will determine if the debris population will rise or fall and the rate at which this change will occur. The balance obviously will vary greatly at different altitudes; in circular orbits below about 500 km, where orbital decay is fairly rapid, major long-term increases in the debris population are unlikely, while at higher altitudes and in some high-eccentricity orbits, medium and large debris added to the population may remain for tens, thousands, or even millions of years.

As described in Chapter 2, models have been developed to predict the future debris population. Many of these models are loosely based on population "birth and death" models, which assign the debris population to a number of "bins," each characterizing the number of particles of a given mass range within a given altitude range. The simplest models (e.g., Talent, 1992) use one bin for all masses and altitudes to represent the entire LEO population, but more complex models have used more mass and altitude ranges (Rossi et al., 1993). NASA's EVOLVE model (Reynolds, 1990) can use a variable number of bins, but typically uses 15 mass and 36 altitude ranges. Although these models incorporate a variety of different assumptions, they generally carry out similar procedures to predict the future debris population.

First, the initial debris population (to a certain limiting size or mass) is calculated for each altitude bin, based on measurements or models of the current debris population. Then the orbits of this initial population are propagated into the future (by using either deterministic or statistical methods), in the course of which some objects are removed from or added to the altitude bin as a result of perturbing forces such as atmospheric drag. Predictions of the amount and distribution of new objects launched into orbit, as well as of the results of possible explosions in orbit, are used to add new space objects to the population. When it is determined that a collision will occur, another model is used to determine the effects of the collision, including the creation of new debris. The entire process is then repeated, with the output of the first iteration used as the initial population for the next.

These models are generally useful only in predicting the magnitude and characteristics of the populations of medium-sized and large debris. As discussed in Chapter 3, knowledge of small debris is so limited that it is extremely difficult to estimate the current population, much less project future trends. About the only predictions that can be made about the future small debris population are that

- the amount of small debris produced in breakups is likely to increase if the number of collisions grows, because collisions are predicted to produce very large numbers of small debris particles; and
- the amount of very small debris (such as the particles expelled by solid rocket motors and the smallest products of breakups and degradation) in orbit may change markedly from year to year, due to the strong effect of perturbing forces on the orbits of these particles. Thus, regardless of the historical total amount of very small debris released into the environment, the population of these particles at any given time in the future will be strongly dependent on the amount produced during the preceding one or two years.

Uncertainties in the Models

As described in Chapter 2, models that predict the future debris population contain numerous uncertainties. These include uncertainties about

- the number, characteristics, and distribution of the objects that are currently in orbit;
- the number, characteristics, and initial distribution of objects that will be launched into orbit in the future;
- the future level of solar activity and its effect on atmospheric drag; and
- the characteristics of the fragments created in space object explosions and collisions.

These uncertainties have varying effects on the accuracy of models that predict the future space object population.

As discussed in Chapter 3, our knowledge of the current population of objects in orbit is very limited; the locations of only the largest are known. If a model of the future debris environment is going to include uncataloged objects in its initial population, it must extrapolate the number and distribution of the uncataloged population by using existing tracking and sampling data as well as estimates of the number and characteristics of debris created during known breakups. (Estimations of the amount of mission-related debris that has been released during spacecraft deployment and operations can also be employed to aid in this process, but they are currently not widely used.) This uncertainty about the uncataloged population, however, has only a limited effect on predictions of the future environment because (as is discussed later) it is the largest objects—most of which are cataloged—that drive the growth in the future population. Some studies show that the untrackable population has no detectable effect on the evolution of the future LEO debris population (Eichler, 1993; Kessler and Loftus, in press). Other models take advantage of this phenomenon and use only cataloged objects for their initial population (Kessler, 1991).

Predictions of the number and characteristics of space objects that will be added to various orbital regions as a result of future launches are also uncertain. The future launch rate, the size ranges of future spacecraft, and the distribution of orbits into which these spacecraft will be launched cannot be predicted in detail because they depend on such unpredictable factors as future mission requirements, technologies, economics, and global politics. Because of the limited predictability of future solar activity (and thus of atmospheric drag in LEO), it is also difficult to estimate the number of objects that will be removed by natural forces.

Finally, breakup models are a major source of uncertainty in any predictive model of the future debris environment. As discussed in Chapter 2, breakup models are based on a very limited set of data, and models of both explosions and collisions contain large uncertainties. Models of collisions, in particular, are supported by data from only two in-space collisions and a few ground tests conducted at velocities lower than typical LEO collision velocities. It is not at all certain how well current collision models

   • incorporate the effects of different space object configurations, different spacecraft materials, and different impact geometries on the results of a collision;
   • approximate the threshold size (in terms of mass and/or energy) of debris that can cause space objects of a given size to break up; or
   • estimate the distribution of the size and mass of particles produced in collisions (although the limited ground and space tests that have been conducted indicate that current models are fairly accurate at predicting the amount of large debris produced in a collision).

Models of the future debris population often deal with these uncertainties by treating them as variables. For example, a model of the future population can be run with the rate at which rocket bodies explode in the future set equal to current levels, and then run again with the rate set to zero, to predict the effects of implementing rocket body passivation measures on the future growth of the space object population. Treating these uncertainties as variables does not reduce the overall uncertainty in the model, of course, but it does serve to clarify the dependence of the model's results on each variable.

Predictions of the Future Orbital Environment

If the only additions to the future debris population were rocket bodies, nonfunctional spacecraft, mission-related debris, and the products of explosions and surface deterioration, the space object population would likely continue its roughly linear growth. Implementation of measures to reduce the number of explosions of spacecraft and rocket bodies, and to limit the amount of mission-related debris released as a result of spacecraft deployment and operations, might result in a slower rate of growth, just as changes in future launch patterns could result in a faster rate of growth. Collisions between space objects, however, threaten to add a potentially large and exponentially growing number of new objects to this population.

The probability that a collision will occur in any particular orbital region increases with roughly the square of the number of objects in that region. Each orbital region has a "critical density," at which point it contains enough objects with sufficient mass that the rate of fragment production from collisions is greater than the rate at which objects are removed due to such forces as atmospheric drag. Once this critical density is reached, fragments from collisions will cause an ever-increasing number of new collisions. This is sometimes referred to as a "cascading effect" or a "chain reaction," although the time frame involved is typically on the order of decades or centuries—not the nearly instantaneous reaction that the latter term often implies. Once collisional cascading has begun, it cannot be stopped by a reduction in launch rate because it is self-sustaining. If no new mass is added to the region, the number of collisions will eventually drop (perhaps over hundreds or thousands of years) as the large objects are broken into smaller pieces, but by that time, the collision hazard in the orbital region may be too high for most space operations.

Several independent models of the future debris population suggest that collisional cascading is likely to occur in Earth orbit (see Kessler and Cour-Palais, 1978; Kessler, 1991; Talent, 1992; Kessler et al., 1993; Rex and Eichler, 1993; Rossi et al., 1993; Su, 1993). Although these simulations use different methodologies and incorporate a number of different values for such parameters as the initial population and the amount of debris produced in collisions, their results are uniform in predicting a more-than-linear increase in the number of future space objects in LEO over the next century unless measures are taken to reduce the addition of new debris to the environment. Figures 8-1 and 8-2 show the results of two of these simulations.

These figures show predictions of the future growth in the population of objects larger than 1 cm, because this population is more relevant to the hazard to spacecraft than the population of large objects. However, it must be remembered that the collisional population growth is driven almost entirely by the population of large debris. The exponential growth visible in predictions of the 1-cm debris population is only a symptom, not the cause, of collisional population growth.

Though these models all show that an exponential rise in the orbital debris population will occur unless preventive measures are taken, the time frame over which this rise will occur cannot be determined precisely. The error bounds of the time frame for collisional growth are a result of all of the uncertainties (discussed earlier) that are incorporated into models of the future population growth. Figures 8-4 and 8-5 show how variations in the assumptions made in a model can affect that model's predictions of the rate at which the future space object population will rise. In Figure 8-4, different assumptions about the basic population growth rate (i.e., the growth rate not counting objects produced in

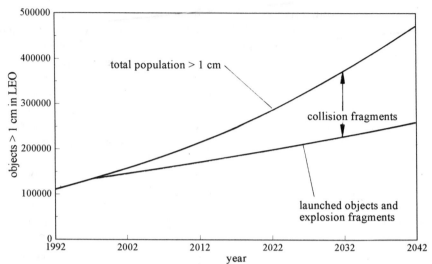

FIGURE 8-1 Model simulation of the contribution of collision fragments to the future LEO space object population. SOURCE: Rex and Eichler, 1993.

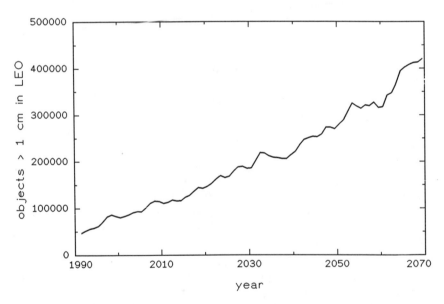

FIGURE 8-2 Different model simulation of the future LEO space object population (constant launch rate and explosion rate at current level, no mitigation measures). SOURCE: National Aeronautics and Space Administration.

BOX 8-1 One Model of the Future Debris Environment

The CHAIN program (Rex and Eichler, 1993) uses a combination of analytic and stochastic methods to predict the future number of orbital objects larger than a centimeter in diameter. Figure 8-3 shows the program's logic flow.

First, the current number of orbiting space objects larger than 1 cm is estimated from measurements of the debris environment and is divided into 24 bins with six mass and four altitude ranges. The collision risk (for 21 different types of collisions) within each altitude bin is then calculated by using analytical formulas. A stochastic Monte Carlo method is used to randomly determine the number of collisions in each altitude bin based on these collision risks. Fragments from these collisions are then added to the population, along with newly launched objects and fragments from space object explosions. The effects of orbital decay on all of these objects are then calculated, and the new "initial" population is determined. The analytic formulas are then applied to this new population to calculate collision probabilities for the next iteration, and the entire process is repeated.

This model predicts that a "business as usual" scenario, in which no debris reduction measures are implemented, would result in the population of objects larger than 1 cm increasng to about 250,000 in the next 50 years—not including the effects of collisions. When the effects of collisions are factored in, the further increase to the population is more than 200,000 additional fragments. Figure 8-1 shows the averaged output of this model.

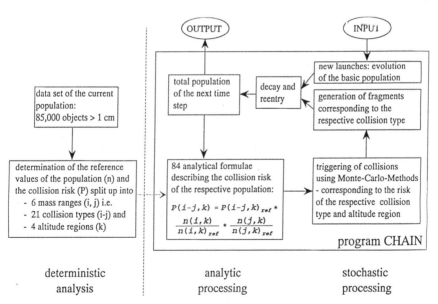

FIGURE 8-3 Logic flow of the CHAIN program. SOURCE: Rex and Eichler, 1993.

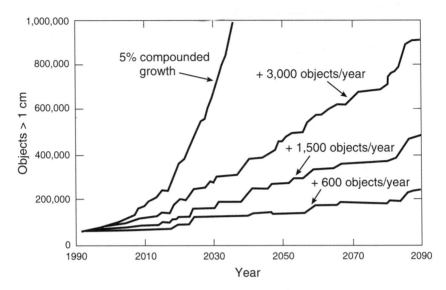

FIGURE 8-4 Effect of basic population growth rate on predicted future population of LEO objects larger than 1 cm. Initial basic population (launched objects plus collision fragments): about 60,000 >1 cm. Lines are the result of Monte Carlo simulations run with the same stochastic numbers but with different basic population growth rates. SOURCE: Eichler et al., 1992.

collisions) change the predicted time it takes for the debris population to double from about 15 to 60 years. Figure 8-5 shows the effect of varying the mass distribution of collision fragments in a different simulation of the future environment; it is clear from the figure that this factor also introduces a large uncertainty into the predicted growth rate of the future debris population. Since models of the future debris population must incorporate both these uncertainties, as well as others, it is premature to suggest exactly when collisional growth will begin to occur; it may already be under way, or it may not begin for several decades.

Collisional growth will not, however, take place over the entire near-Earth orbital area. It is most likely to occur in regions that (1) have a high debris flux, (2) do not experience a high level of atmospheric drag, and (3) have high typical collision velocities. (These characteristics lead to a large number of energetic collisions that produce long-lasting fragments.) Figure 8-6 shows one assessment of how the critical density varies with altitude in LEO due to such factors as atmospheric drag and the size and inclination distribution of the current populations (Kessler, 1991). The shaded area shows the two LEO regions, at 900 to 1,000-km and around

1,500-km altitude, in which the cataloged population has already exceeded the calculated critical density.

In regions where the spatial density has already exceeded the critical density, the number of collision fragments produced will eventually rise exponentially. The launches of new spacecraft (and their accompanying rocket bodies and mission-related debris), as well as the explosion-induced breakup of orbiting rocket bodies and spacecraft, are also likely to contribute to the debris population in these regions. Launches to these regions will probably continue for some time because orbits within these regions have characteristics that make them valuable to spacecraft operators; this is, after all, the reason they became crowded. As discussed in Chapter 7, some level of residual explosions is also likely to continue, regardless of the passivation measures adopted. The addition of new objects to already crowded orbital regions will likely increase the collision probability for functional spacecraft in these regions, as well as the rate and magnitude of their future collisional population growth.

LEO regions that have not reached a critical density may still be af-

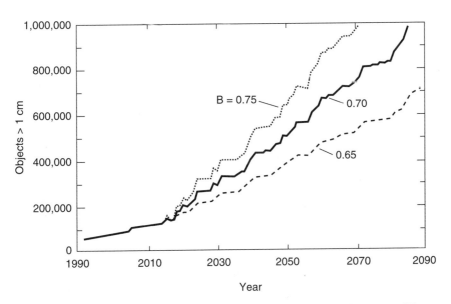

FIGURE 8-5 Effect of assumed mass distribution of collision fragments (*B*) on prediction of future population of objects larger than 1 cm. The three scenarios are the result of Monte Carlo simulations run with the same stochastic numbers but with different assumed mass distributions of collision fragments. In all scenarios, the basic population (launched objects plus collision fragments) increases by 3,000 objects >1 cm per year. SOURCE: Eichler, 1993.

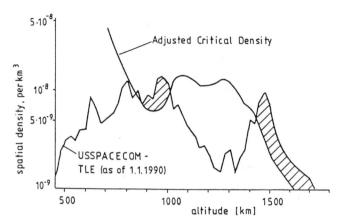

FIGURE 8-6   Critical density in LEO (shaded regions are above the adjusted critical density). SOURCE: Kessler, 1991.

fected by collisions in those regions that have surpassed their critical density. For example, as can be seen from Figure 8-6, the debris population at altitudes below about 700 km is not likely to exceed the critical density; at these altitudes, atmospheric drag typically will remove collision fragments before they collide with another object. Collision fragments produced in other orbital regions, however, could increase the debris hazard in these low-altitude orbits. Although most of the larger debris generated in collisions in higher orbits will tend to stay relatively close to their initial orbit, some smaller fragments will typically be ejected at greater velocities (as discussed in Chapter 4) and thus will be distributed widely, increasing collision probabilities over the entire LEO region. In addition, fragments from collisions at higher altitudes will eventually experience orbital decay, causing them to pass through lower orbital regions. Finally, fragments produced in collisions between objects in highly elliptical orbits and objects in LEO regions with high spatial densities may pass through other LEO regions at high velocities.

Even though atmospheric drag is only a minor factor in removing debris from orbits above LEO, the chance of collisional cascading does not necessarily increase with altitude as might be expected. There are a number of reasons for this. First, as shown in Figure 3-2, higher orbits typically have a much lower object flux than LEO. Second, the volume of a given altitude region increases with altitude so, even if additional objects are added, the spatial density will rise only slowly. Third, collision velocities in high-altitude orbits are generally slower than in LEO; this typically leads to the creation of fewer fragments in a collision. Finally (as discussed in Chapter 7), the debris created in collisions at high altitudes is dispersed over a much wider area than debris from collisions in LEO, which reduces its chance of striking other objects in the initial orbital region.

In GEO, where collision velocities are lower than in LEO and debris produced in collisions will be spread over a wide volume of space, collisional cascading may not occur and, if it does, may not be noticeable for thousands to tens of thousands of years (Kessler, 1993). It thus seems probable that unlike the situation in LEO, the future debris population in GEO could be driven more by explosions and continued launch traffic than by collisions. Because of this dependence on inherently unpredictable factors, it is difficult to make accurate predictions of the future GEO population.

EFFECTIVENESS OF DEBRIS REDUCTION MEASURES

There are two major types of methods to reduce the debris hazard in Earth orbit. One set of methods aims at limiting the *number* of potentially harmful objects in orbit, primarily by reducing the release of mission-related debris and by preventing further explosions of spacecraft and rocket bodies. The other methods seek to limit the total *mass* and *cross-sectional area* of objects in orbit by deorbiting or reorbiting spacecraft and rocket bodies. These two approaches to debris reduction have very different effects on the short- and long-term orbital debris hazard.

Limiting the number of potentially harmful objects in orbit can sharply reduce growth in the short-term debris hazard and can restrict growth in the long-term hazard to some regions, but it will have little effect on slowing or preventing collisional cascading. Limiting the total mass and cross-sectional area added to orbit, on the other hand, can prevent or slow the onset and growth of collisional cascading and can also ameliorate the short-term collision hazard. Limiting cross-sectional area plays an important role in reducing the long-term potential for collisional cascading because the total cross-sectional area in orbit represents the total "target area" for collisions. Limiting the amount of mass is important because, in the long-term, the mass in orbit determines the maximum number of collision fragments capable of causing further breakups.

The same models used to predict the future evolution of the debris population can also be used to predict the effectiveness of various measures in limiting the growth in that population. Although all of the uncertainties in these models (discussed earlier in this chapter) also apply to such predictions, that does not prevent a rough assessment of the effectiveness of various debris reduction methods from being made. These

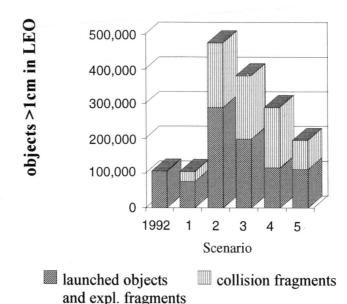

FIGURE 8-7 Predicted effect of debris reduction measures on population of collision fragments and other debris in 2042. SOURCE: Rex and Eichler, 1993.

models simulate only the LEO environment, however; similar models with the capability to predict the effectiveness of debris reduction measures on the semisynchronous or GEO environment have not yet been developed.

Figures 8-7 and 8-8 show the result of one of these model simulations (Rex and Eichler, 1993). Figure 8-7 shows the predicted effect of various debris reduction methods on the LEO population of collision fragments and other types of debris in 2042, and Figure 8-8 shows how these measures are predicted to affect the total LEO debris population over time. In Figure 8-7, the first bar shows the estimated population of LEO objects larger than 1 cm in 1992, and the other bars show the predicted 2042 population of debris if various reduction methods are implemented. Scenario 1 shows the predicted 2042 LEO population resulting from an immediate and complete cessation of all space launches. The model suggests that in this case, although the population of launched objects and explosion debris would decrease as a result of orbital decay, some collision debris and some debris from explosions of objects already in orbit would be generated, keeping the total population nearly constant over the 50 years of the simulation. Scenario 2 represents the other extreme, in which the linear growth of space activity drives the population of objects

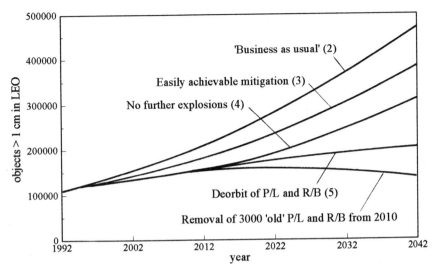

FIGURE 8-8 Predicted effect of debris reduction measures over fifty years on space object population. SOURCE: Rex and Eichler, 1993.

larger than 1 cm in LEO from about 100,000 to close to 500,000 over the 50 years of the simulation.

Scenarios 3, 4, and 5 explore the effects of various debris reduction measures on the future LEO population. Scenario 3 shows the effect of cutting the explosion rate and the release of mission-related debris in half, and scenario 4 shows the effect of completely ending all explosions in orbit (starting in 1998). These measures are shown to significantly cut the predicted number of objects added to orbit, but since they do nothing to reduce the total mass or cross-sectional area added to orbit, it is not surprising that the model predicts that they will not greatly reduce the number of additional collision fragments. Figure 8-8 shows that neither scenario 3 nor scenario 4 prevents the eventual exponential growth of the debris population.

Scenario 5 shows the combined effect of preventing explosions, reducing the release of mission-related debris, and performing EOL deorbiting of rocket bodies (after 2003) and spacecraft (after 2010). Even in this scenario, the total debris population still doubles in 50 years. In this case, however, unlike the previous cases, the population does not increase rapidly near the end of the simulation, which suggests that exponential growth has at least been delayed. Finally, the bottom curve in Figure 8-8 shows how (for this model) it would take the active removal of 3,000 old payloads and rocket bodies to actually prevent the population from growing.

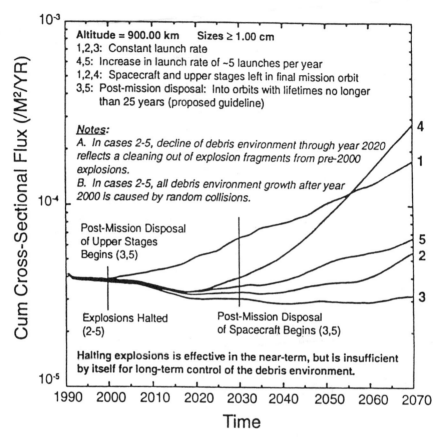

FIGURE 8-9 Predicted effect of debris mitigation measures on the debris environment. SOURCE: Loftus, 1993.

Figures 8-9 and 8-10 show the results of a different model's predictions of the effect of debris reduction measures (Loftus, 1993). This model incorporates different initial assumptions than the previous model and examines the effect of a different set of potential debris reduction methods. The figures focus on the object flux in the 900-km-altitude regime, which (as discussed earlier) is believed to already have a spatial density that exceeds the region's critical density. Like the previous model, this model predicts that the debris population will rise much more rapidly if the number of explosions is not reduced and that the debris population will rise exponentially unless some types of reorbiting maneuvers are performed. (Figure 8-9 also shows the strong influence of the assumptions about the future launch rate on the future hazard; scenario 4, which assumes that the number of launches increases from the present level at a rate of five additional launches each year, predicts double the flux in 2070 than does scenario 2, which is identical except that it assumes a constant launch rate at current levels.) Figure 8-10 further explores the effect on the debris flux of reducing the orbital lifetime of spacecraft and rocket bodies. It is clear from this figure that orbital lifetime reduction measures can slow the rate at which the debris population increases and that the greater the reduction in orbital lifetime, the less will the debris environment grow. However, given all the uncertainties involved, the model does not suggest that any one particular chosen target lifetime is more cost-effective than another.

It must be remembered that all of these models incorporate high levels of uncertainty. This can be seen just by comparing Figure 8-8 and Figure 8-9. In Figure 8-9, the debris hazard for a "no-explosion" scenario

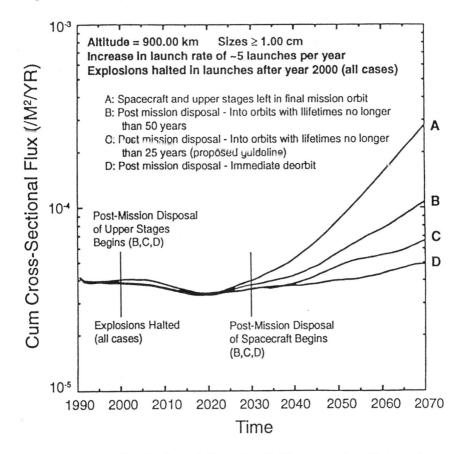

FIGURE 8-10 Predicted effect of disposal orbit lifetime on the debris environment. SOURCE: Loftus, 1993.

with a linear traffic growth actually declines until about 2020. In the equivalent scenario (4) in Figure 8-8, the population almost doubles over the same time frame. Detailed comparisons of the two models (Reynolds and Eichler, in press) indicate that the differences are mostly due to different assumed initial conditions for the amount of mass in orbit and, to a lesser extent, the number of bins used. Nevertheless, both models predict that (1) reducing the number of explosions in orbit will help contain the growth in the debris hazard, and (2) deorbiting or lifetime reduction maneuvers (in addition to reducing the number of explosions) may be required to prevent an eventual exponential rise in the debris population.

FINDINGS

Finding 1: If the only additions to the future debris population were rocket bodies, nonfunctional spacecraft, mission-related debris, and the products of explosions and surface deterioration, the space object population would probably continue its roughly linear growth. Several models using different methodologies and different assumptions, however, predict that collisions between space objects will add a potentially large and exponentially growing number of new objects to this population. Because of the numerous uncertainties involved in models of the debris environment, it is premature to suggest exactly when collisional growth will begin to occur; it may already be under way, or it may not begin for several decades.

Finding 2: Collisional growth is most likely to occur in regions that (1) have a high debris flux, (2) do not experience a high level of atmospheric drag, and (3) have high typical collision velocities. Two LEO regions that meet these criteria, at around 900- to 1,000-km and around 1,500-km altitude, are believed to already have exceeded their critical density, the point at which they will continue to experience population growth due to collisions even if no further objects are added. Fragments from collisions in regions experiencing collisional growth may be widely distributed, increasing the collision probability even in regions that are not threatened by collisional growth.

Finding 3: Although debris fragments represent the greatest short-term debris hazard to current spacecraft, it is the large objects in orbit (generally spacecraft and rocket bodies) that drive the potential for collisional cascading. Thus, although eliminating the explosion of objects in orbit can reduce the short-term growth in the debris population, it is necessary to take measures to remove spacecraft and rocket bodies from crowded orbital regions at the end of their functional lifetimes in order to reduce the potential for collisional growth.

REFERENCES

Eichler, P. 1993. The necessity and efficiency of mitigation measures to limit the debris population in LEO. Presented at the National Research Council Committee on Space Debris Workshop, Irvine, California. November 18.

Eichler, P., H. Sdunnus, and J. Zhang. 1992. Reliability of Space Debris Modelling and the Impact on Current and Future Space Flight Activities. World Space Congress paper B.8-M.3.11. Washington, D.C.

Kessler, D.J. 1991. Collisional cascading: The limits of population growth in low Earth orbit. Advances in Space Research 11(12):63–66.

Kessler, D.J. 1993. Orbital debris environment. Pp. 251–262 in Proceedings of the First European Conference on Space Debris, Darmstadt, Germany, 5–7 April 1993. Darmstadt: European Space Operations Center.

Kessler, D.J., and B.G. Cour-Palais. 1978. Collision frequency of artificial satellites: The creation of a debris belt. Journal of Geophysical Research 83(A6):2637–2646.

Kessler, D.J., and J.P. Loftus, Jr. In press. Orbital Debris as an Energy Management Problem. Paper presented at the 31st Plenary Meeting of COSPAR, Hamburg, Germany, July 14. To be published in Advances in Space Research.

Kessler, D.J., R.C. Reynolds, and P.D. Anz-Meador. 1993. Review of Various Models to Describe the Orbital Debris Environment. IAA-6.3-93-744. Paper presented at the 44th Congress of the Astronautical Federation, Graz, Austria, October 16–22.

Loftus, J.P. 1993. The NASA debris handbook. Presented at the National Research Council Committee on Space Debris Workshop, Irvine, California. November 18.

Rex, D., and P. Eichler. 1993. The possible long term overcrowding of LEO and the necessity and effectiveness of debris mitigation measures. Pp. 604–615 in Proceedings of the First European Conference on Space Debris, Darmstadt, Germany, 5–7 April 1993. Darmstadt: European Space Operations Center.

Reynolds, R. 1990. A review of orbital debris environment modeling at NASA/JSC. AIAA 90-1355. Paper presented at the AIAA/NASA/DoD Orbital Debris Conference, Baltimore, Maryland, April 16–19. Washington, D.C.: American Institute of Aeronautics and Astronautics.

Reynolds, R., and P. Eichler. In press. A comparison of debris environment projections using the EVOLVE and CHAIN models. Paper presented at the 31st Plenary Meeting of COSPAR, Hamburg, Germany, July 14. To be published in Advances in Space Research.

Rossi, A., L. Anselmo, C. Pardini, A. Cordelli, P. Fannella, and T. Parinello. 1993. Approaching the exponential growth: Parameter sensitivity of the debris evolution. Pp. 287–289 in Proceedings of the First European Conference on Space Debris, Darmstadt, Germany, 5–7 April 1993. Darmstadt: European Space Operations Center.

Su, S.-Y. 1993. On runaway conditions of orbital debris environment. Advances in Space Research 13(8):221–224.

Talent, D. 1992. Analytical model for orbital debris environment management. Journal of Spacecraft and Rockets 29(4):508–513.

9

Recommendations

IMPROVING KNOWLEDGE OF THE DEBRIS ENVIRONMENT

Understanding the orbital debris environment (including debris size ranges, compositions, and distribution by orbital altitude, eccentricity, and inclination) is necessary to assess the debris hazard to spacecraft in various orbits, to understand the future evolution of the debris population, and to enable wise decisions to be made on methods to reduce the future hazard. However, data are lacking on many debris sources, size ranges, and orbital regions; current understanding of the debris environment is based on incomplete measurements and models that are not yet mature.

Increasing our knowledge of the orbital debris environment and applying that knowledge to debris mitigation practices may be the most cost-effective means of reducing the future impact of the debris hazard. First, better understanding of the environment would help spacecraft designers to protect spacecraft more effectively against debris. Although some meaningful measurements have been made at lower altitudes, current understanding of the debris environment is not sufficient for most spacecraft designers to predict accurately the level of debris protection that spacecraft may require; this may result in costly over- or underprotection. Second, a better understanding of the environment could be applied to determine which debris prevention measures will most effectively reduce the future hazard. Currently, there is much uncertainty about the cost-effectiveness of some methods of reducing the future debris hazard; models of the future debris population incorporating new data (e.g., data on previously uncataloged large debris) could help to identify the best methods with which to deal with the problem and the orbital regions at which these methods should be targeted.

This is not to suggest an effort to characterize all debris in all orbits; rather, characterization efforts should focus on gathering the information needed to fill critical data gaps. Previously, most measurements of the debris environment were made when opportunities arose. Although these measurements added greatly to our knowledge of the debris environment, and further ad hoc measurements will doubtless continue to add to our knowledge, future debris characterization efforts should focus on either (1) providing information that will be directly useful to spacecraft designers and operators, or (2) answering questions about the debris environment that will increase understanding of the population's long-term evolution.

Currently, the only national or international guidance on either the most important areas in the debris field to be investigated or potential methods to investigate these areas comes from the Inter-Agency Space Debris Coordination Committee (IADC), which is made up of representatives from ESA, the Russian Space Agency, space agencies from Japan, and NASA. To provide future guidance for debris research, the committee recommends the following:

Recommendation 1: An expanded international group should be formed to advise the space community about areas in the orbital debris field needing further investigation and to suggest potential investigation methods. This group, which could include representatives from industry and academia, as well as from governments, could build on the work of the IADC. The group could identify the highest-priority areas of interest to orbital debris researchers and spacecraft operators, the data required to understand each area, and potential methods to acquire the data.

The committee recommends the following as an interim set of debris characterization research priorities:

Recommendation 2a: Models of the future debris environment should be further improved by refining theoretical models, acquiring and incorporating new data to lessen uncertainties, and testing the models against new data. Ensuring that these models incorporate all major sources of debris and increasing the accuracy of breakup models (for both collisions and explosions) should be major components of this effort. Improving these models is crucial because potentially very expensive decisions on the adoption of debris mitigation measures depend on their conclusions. These decisions must be based on the best information possible.

Recommendation 2b: *Uncataloged medium-sized and large debris in LEO should be carefully studied. This should include a long-term measurement campaign to understand more fully the fluctuations in the uncataloged population due to perturbation forces and various generation mechanisms, thorough processing of the data, etc.* Although the composition and dynamics of cataloged debris have been studied fairly well, knowledge of uncataloged large and medium-sized debris is limited. Although uncataloged large and medium-sized debris will not contribute significantly to collisional population growth, this population of debris is more likely to cause significant damage to typical spacecraft than the populations of either cataloged debris or small debris.

Recommendation 2c: *Further studies (including both measurements and modeling) should be conducted to better understand the GEO debris environment. These should include efforts to determine the current debris population in GEO as well as to model its future evolution.* Data on the debris environment in GEO are extremely sparse. Although the chance of a damaging impact in GEO is likely to be much lower than the chance in LEO, it is important to better understand the GEO debris environment because (1) the geostationary orbit is a limited and valuable resource that should be preserved for the future, (2) the orbital lifetime of space objects in GEO is extremely long (on the order of tens of thousands to millions of years), and (3) there are currently many highly valuable spacecraft in GEO.

Recommendation 2d: *A strategy should be developed to gain a better understanding of the sources and evolution of the small debris population.* Because the population of small debris is so time dependent, this strategy should focus on answering questions about the long-term nature of this population. The orbital debris community (including experts in modeling, detection and tracking, impact damage, and damage mitigation) should develop a strategy of observing requirements to effectively provide information about the sources and evolution of the small debris population.

Recommendation 2e: *The data acquired from continuing studies of the debris environment should be compiled into a standard population characterization reference model.* Methods should be adopted to validate or indicate the state of validation of this model. Such a model would aid experimenters in properly interpreting their data and spacecraft designers in properly assessing the hazard to their spacecraft.

In addition, the committee recommends two measures to improve the efficiency and accuracy of research on orbital debris:

Recommendation 3: *The creation of an international system for collecting, storing, and distributing data on orbital debris should be explored. This would include the creation of a unified database and catalog of debris that would receive measurements from all sensors gathering data on debris (including those in the SSS and the SSN). The information from this database would be accessible to interested parties under certain conditions.* Currently, there is no formal mechanism allowing the nations of the world that engage in space monitoring to share data. Sharing of data from the SSS and the SSN could increase confidence in the catalogs of large debris and would also be useful in determining the desirability of future collaborative space debris monitoring efforts. The distribution of data from other sensors would enable an expanded group of researchers worldwide to analyze orbital debris data.

Recommendation 4: The orbital debris community should exercise more peer review over its research. Orbital debris is sometimes studied with an eclectic and often not fully developed set of observational, experimental, and modeled data and methods. The field needs a more rigorous scientific structure to give it a better theoretical underpinning and to logically link its elements. The practices of using external technical peer review panels, publishing in peer-reviewed journals, and establishing a close working relationship with related scientific fields should be expanded to provide some of this rigor.

IMPROVING SPACECRAFT PROTECTION AGAINST DEBRIS

Even if fairly drastic steps are taken to reduce the generation of new debris, a hazard will likely continue to exist, and probably grow, in some important orbital regions for a great many years. Without remedial steps, the debris hazard will grow more rapidly. In either case, orbital debris is now a part of the space environment and should be considered during the design of spacecraft and the planning of space operations. As described in Chapter 6, the growing availability of (1) analytic and experimental tools to quantify the debris threat to spacecraft and (2) techniques to protect against debris impacts make it feasible for designers to assess the debris hazard and protect spacecraft appropriately. However, not all spacecraft designers have knowledge of these tools and techniques. For this reason, the committee recommends the following:

Recommendation 5: A guide to aid spacecraft designers in dealing with the debris environment should be developed and distributed widely. This design guide should include information on environmental prediction, damage assessment, and passive and operational protection techniques. Such a guide would enable spacecraft designers (1) to assess the need to incorporate protective measures in spacecraft design or operations and (2) to choose and implement appropriate measures of protection, if necessary. It could also serve as a useful reference text for advanced students in space engineering. Chapter 6 provides a top-level description of the processes and methods that should be discussed in this guide.

Considerable effort has already been invested in studying the effects of debris impact on spacecraft and the ability of shielding to reduce impact damage. However, as discussed in Chapter 5, knowledge gaps remain in (1) the effects of impact by the variety of debris shapes and compositions likely to exist in orbit, (2) the vulnerability of different spacecraft components to debris impact, and (3) the effects of impact at typical LEO collision velocities. To better predict impact damage and design debris shields, the committee recommends the following:

Recommendation 6: *Research should be continued to characterize the effects of hypervelocity impacts on spacecraft systems in the following areas:*

- *further development of techniques to launch projectiles to the velocities typical of collisions in LEO;*
- *improved models of the properties of newer spacecraft materials.*
- *studies of damage effects on critical components;*
- *development of analytical tools consistent over a range of debris impact velocities, shapes, and compositions; and*
- *improved models of catastrophic spacecraft breakup from debris impact.*

The first four of these research areas aim at improving spacecraft and shield design; the final research area aims at improving models of the future debris population.

These research goals could be achieved more easily if data from hypervelocity facilities worldwide were made more readily available. Unfortunately, as discussed in Chapter 5, the capabilities of many hypervelocity facilities are not well known, and the impact data generated at these facilities are often inaccessible. This has resulted in duplication of effort both within and between nations, slowing the development of good models of debris impact damage. Thus, the committee recommends the following:

Recommendation 7: *A handbook describing the capabilities of the international hypervelocity impact facilities generally available for debris research should be developed.* Such a handbook would facilitate the sharing of impact results generated at different facilities, perhaps leading to the establishment of a debris-related database of impact results accessible via the Internet.

REDUCING THE FUTURE DEBRIS HAZARD

Unless the production of new debris is reduced, it will become necessary for increasing numbers of spacecraft to adopt measures to avoid debris impact damage, and the chance of losing functional spacecraft to debris will increase. As discussed in Chapter 7, cleaning up debris via active removal will be uneconomical for the foreseeable future, so efforts must focus on reducing the creation of new debris. There are many possible means of accomplishing this goal, but the decision on which should be implemented cannot be made solely on technical grounds. As with other environmental issues, decisions on the adoption of debris reduction methods, and on the means to implement these methods, must balance political and economic as well as technical factors and thus must be made in forums that are capable of balancing all of these factors.

Current international law does not specifically address the orbital debris issue, so there is a fairly clean slate upon which to draft future regulations to reduce the generation of new debris. (Existing international agreements pertaining to orbital debris, as well as some of the efforts under way that may affect future rule making on orbital debris-related issues, are discussed briefly in Appendix A.) Possible future regulatory schemes may be voluntary or mandatory; they may provide incentives to spacecraft operators who reduce debris creation, or they may specify particular debris-mitigating measures all manufacturers must incorporate. It is clear, however, that debris reduction measures enacted by any single nation will not be sufficient to prevent a growing future hazard. For this reason, and because unilaterally adopted debris reduction measures may reduce economic competitiveness, the committee recommends the following:

Recommendation 8: The spacefaring nations should develop and implement debris reduction methods on a multilateral basis. Given the long development cycle for new space vehicles with debris-minimizing features, the technical development, cost-benefit assessments, and international discussion required to implement countermeasures should start as soon as possible.

Although these multilateral discussions cannot be conducted on a purely technical basis, it is crucial that they be based on sound technical advice. The committee's consensus technical assessment of the actions that should be implemented to reduce future growth in the debris hazard, based on its current understanding of the debris environment and of the costs and benefits of various mitigation measures, is represented in the following recommendations (Chapters 7 and 8 discuss each of these actions in greater detail):

Recommendation 9: *Space system developers should adopt design require-ments ensuring that spacecraft or rocket bodies do not explode after their func-tional lifetimes.* Ensuring that all potential sources of stored energy on a spacecraft or rocket body are depleted at the end of their functional life-time is the primary means of accomplishing this goal. Explosions of spacecraft and rocket bodies have been major contributors to the debris hazard, so preventing such explosions will significantly reduce the growth in the short-term debris hazard. Implementing design features to passivate spacecraft and rocket bodies after their functional lifetimes will generally not be very costly.

Recommendation 10: *The release of mission-related objects during spacecraft deployment and operations should be avoided whenever possible. Release of mission-related objects in long-lifetime orbits should be particularly avoided.* Mission-related debris is a significant fraction of the population of large debris in orbit. Reducing the release of mission-related debris during spacecraft deployment and operations can typically be accomplished without significant expenditure and, in general, without new technology, although some hardware development may be required.

Recommendation 11: *Developers should incorporate requirements that space-craft and rocket bodies be designed to minimize the unintentional release of surface materials, including paint and other thermal control materials, both dur-ing and after their functional lifetimes. To aid in meeting these requirements, surface materials that minimize the release of small particles should be developed and used.* The deterioration of spacecraft surfaces (paint, etc.) is believed to be a major contributor to the population of small debris, so ending its release would prove beneficial to the space environment.

Recommendation 12: *Intentional breakups in orbit (especially those expected to produce a large amount of debris) should be avoided if at all possible. No intentional breakups expected to produce numerous debris with orbital lifetimes longer than a few years should be conducted in Earth orbit.* Occasionally, an organization may want to explode a space object in orbit for defense, scientific, or calibration purposes. If it is absolutely necessary that the breakup take place in Earth orbit, it should be at a low altitude to limit the maximum orbital lifetime of fragments.

All of these actions will help to reduce the short-term debris hazard, but (as described in Chapter 8), models of the future debris population show that EOL reorbiting of large objects (generally rocket bodies and spacecraft) in LEO or in orbits that pass through LEO may be necessary to reduce collisional growth in the LEO debris population. However, removing these objects from orbit (particularly from the higher orbits) can be costly. Ensuring that spacecraft and rocket bodies passing through LEO are (after their functional lifetime) placed into orbits that will decay in a reasonable amount of time appears to be the most cost-effective reorbiting measure. Determining exactly how long this time should be will be as much a political and economic decision as a technical one, due to the relatively large costs that such a maneuver may impose on some missions. Because of the long lead time required to develop and qualify new space hardware, however, it is necessary to begin setting standards now. For this reason, the committee recommends the following:

Recommendation 13: Spacecraft and rocket bodies in LEO and in highly elliptical orbits passing through LEO should be reorbited after their functional lifetime. This reorbiting maneuver should either remove them from LEO or reduce their orbital lifetime. Effort should be made to achieve an international consensus on the magnitude of such reorbiting maneuvers. A draft NASA guideline suggested that spacecraft in orbits that pass through LEO should be limited to orbital lifetimes in LEO of no longer than 25 years after mission completion; this standard does not seem unreasonable. However, any orbital lifetime limitation guideline that is adopted should be based on thorough scientific analysis.

Although the geosynchronous region may not be subject to collisional cascading and current GEO hazard levels from orbital debris appear to be very low, the hazard from debris left in GEO can persist for millennia. Currently, the long-term evolution of the debris environment is not well enough understood to determine the best long-term strategy for managing the debris hazard in GEO. Experts have not yet reached a consensus on the best locations for disposal orbits, or even on whether the use of disposal orbits is the optimal strategy for containing the GEO debris hazard. However, it may not be wise to let the GEO debris population grow until a permanent solution is divined. For these reasons, the committee recommends the following:

Recommendation 14: The use of GEO disposal orbits should be further studied. Until such studies produce a verifiably superior long-term strategy for dealing with the GEO hazard, operators of GEO spacecraft and rocket bodies should be encouraged to reorbit their spacecraft at EOL if they are capable of safely performing a reorbiting maneuver to a disposal orbit at least 300 km from GEO. Studies on the use of GEO disposal orbits should be focused on the development of a long-term strategy for maintaining a low debris hazard in GEO. Such studies should include the development of accurate models capable of predicting the effects of various debris reduction measures on the future hazard in GEO.

APPENDIXES

A

Space Law and Orbital Debris

As mentioned in Chapter 9, orbital debris is not addressed explicitly in current international law. International agreements that directly address orbital debris, however, may eventually be needed to deal with a number of debris-related issues. This appendix briefly summarizes some of the existing space law potentially applicable to the debris issue and discusses some of the activities currently under way that may affect future international rule making on orbital debris-related issues. More detailed discussions of the legal regime and its application to the debris issue are contained in the references listed at the end of this appendix.

United Nations (UN) Treaties

In the past, international space laws have been created under the auspices of the UN Committee on the Peaceful Uses of Outer Space (COPUOS). To date, three treaties with potential relevance to orbital debris issues have entered into force:

- the Treaty on Principles Governing the Activities of States in the Exploration and Use of Outer Space, Including the Moon and Other Celestial Bodies, October 10, 1967 (the Outer Space Treaty);
- the Convention on International Liability for Damage Caused by Space Objects, September 1, 1972 (the Liability Convention); and
- the Convention on Registration of Objects Launched into Outer Space, September 15, 1976 (the Registration Convention).

Three articles in the Outer Space Treaty contain language pertinent to orbital debris issues. Article VI declares, "States party to this treaty shall bear international responsibility for national activities in outer space." Article VII makes states party to the treaty internationally liable for damage caused by objects (and the component parts of those objects) that they launch or have launched into space. Finally, Article IX allows states that have reason to believe that a planned activity or experiment would cause potentially harmful interference with other space activities to "request consultation" concerning the activity or experiment.

The Liability and Registration Conventions further explore the liability of states for damage caused by their space objects. The Liability Convention makes states liable for damage "caused elsewhere than on the surface of the Earth to a space object of one launching state or to persons or property on board such a space object of another launching state . . . only if the damage is due to its fault or the fault of persons for whom it is responsible." The Registration Convention seeks to provide information for use in determining liability by mandating that all launching states notify the UN of any objects they launch and provide the UN with the objects' orbital parameters. Article VI of the Registration Convention directs nations with monitoring or tracking facilities to aid in the identification of space objects that caused damage.

Although these three UN treaties deal with some of the issues raised by the presence of orbital debris, many other debris-related issues are not addressed. For example, the treaties do not address the potential need for measures to reduce the creation of new debris. (The only reference that may be applicable is Article IX of the Outer Space Treaty, which calls for "consultations" if member states believe activities or experiments would cause potentially harmful interference with other space activities.) In addition, some of the issues that are raised in the treaties are difficult to apply to debris. For example, the liability convention assigns liability based on ownership of the objects involved, but the origin of the vast majority of debris objects that are not cataloged cannot be determined. Even where the treaties may be applicable to debris issues, interpretation is often difficult because the legal definitions of "space debris" and "space objects" are not entirely clear.

Expectations still exist that the UN may eventually create formal rules regarding the creation of orbital debris. The issue of orbital debris has not yet been treated in the COPUOS Legal Subcommittee, but in February 1994, the UN General Assembly made orbital debris a formal agenda item for the COPUOS Scientific and Technical Subcommittee. During that session, the subject of orbital debris was addressed by many national delegations and a number of technical papers were presented. At the session, some delegations advocated that space debris should also be treated in the Legal Subcommittee, but other delegations considered such an action to be premature.

Activities That May Influence Future Orbital Debris Regulations

A number of activities outside the UN may affect future laws and policies on orbital debris issues. These include efforts by such organizations as the International Telecommunication Union, the IAA, the International Law Association, the IADC, and others. Three of these efforts are further detailed below:

International Law Association (ILA)

The Space Law Committee of the ILA has studied legal aspects of orbital debris since 1986. In August 1994, the ILA adopted (in a resolution) a draft "International Instrument on the Protection of the Environment from Damage Caused by Space Debris." This instrument, structured in 16 articles, is the first legal text on space debris agreed to by an international body. It contains a definition of space debris, describes the general obligation of states and international organizations to cooperate (inform, consult, and negotiate in good faith) in the prevention of damage to the space environment. Although this instrument does not constitute law or policy and does not address the technical means to reduce the creation of orbital debris, it could potentially serve as a first step in moving the debris issue into the legal regime.

Inter-Agency Space Debris Coordination Committee (IADC)

Interagency orbital debris coordination meetings involving the ESA, the Russian Space Agency, NASA, and the space agencies of Japan are held biannually. Though these meetings do not deal with the legal aspects of the orbital debris issue, the technical issues of space debris measuring, modeling, and reduction techniques are discussed in detail. Since the four attending space agencies are involved in the majority of all space activities, these meetings represent the biggest forum of practical expertise in the field, and may help to provide the sound technical background needed for the development of any new legal rules on debris.

International Astronautical Academy (IAA)

The IAA issued a "Position Paper on Orbital Debris" in October of 1993. This paper was written by an Ad hoc Expert Group of the IAA's

Committee on Safety, Rescue, and Quality, and was reviewed by that committee and by members of other IAA committees and of the International Institute of Space Law before being approved by the IAA board of trustees and published in *Acta Astronautica*. The position paper contains a brief technical discussion of the present and future debris situation and suggests a number of debris control measures ranked by priority. As the output of an international body of debris experts, this paper may influence future regulations on orbital debris.

National and Regional Policies on Orbital Debris

National laws or policies on orbital debris may potentially affect not only domestic space activities but also any international rule making on the debris issue:

• In the United States, current policy (issued in 1988 by President Reagan) states that "all space sectors will seek to minimize the creation of space debris . . . consistent with mission requirements and cost effectiveness." Another U.S. initiative is NASA's "Space Debris Handbook," which may become an important technical reference for space debris reduction measures.

• The Russian Federation also has a policy on debris, alluded to in Section I, Article 4, Paragraph 2 of its Law on Space: "For the purpose of ensuring strategic and ecological safety in the Russian Federation, the following are forbidden: . . . harmful pollution of space, leading to unfavorable environmental changes, including intentional destruction of space objects in space."

• ESA has had specific requirements to prevent the creation of new debris since 1988. In 1989 ESA's Council passed a resolution defining the agency's objectives in the field of space debris. ESA's policy is ". . . to reduce to the maximum possible extent the production of space debris and to promote exchange of information and cooperation with other space operators . . ."

BACKGROUND REFERENCES

AIAA (American Institute of Aeronautics and Astronautics). 1992. Orbital Debris Mitigation Techniques: Technical, Economic, and Legal Aspects. SP-016-1992. Washington, D.C.: AIAA.

Baker, H.A. 1989. Space Debris: Legal and Policy Implications. Norwell, Massachusetts: Kluwer Academic Publishers.

Baker, H.A. 1994. Regulation of orbital debris—Current status. Pp. 180–188 in Preservation of Near-Earth Space for Future Generations, Simpson, J.A., ed. New York: Cambridge University Press.

Christol, C.Q. 1982. The Modern International Law of Outer Space. New York: Pergamon Press.

Christol, C.Q. 1993. Scientific and Legal Aspects of Space Debris. Paper presented at the 44th Congress of the International Astronautical Federation, Graz, Austria: October 16–22.

International Academy of Astronautics Committee on Safety, Rescue, and Quality. 1992. Position Paper on Orbital Debris. August 27. Paris: International Academy of Astronautics.

U. S. Office of Technology Assessment. 1990. Orbiting Debris: A Space Environmental Problem. Background Paper OTA-BP-ISC-72. September. Washington, D.C.: U.S. Government Printing Office.

B

Workshop on Space Debris

The Committee on Space Debris held a workshop at the Beckman Center of the National Academies of Sciences and Engineering in Irvine, California, on November 18-20, 1993. The following participants attended and provided a great deal of input to the committee on a wide range of debris-related topics.

Invited Participants

Mr. Howard Baker, Department of Justice, Canada

Dr. Vladimir Chobotov, The Aerospace Corporation, California, United States

Mr. Eric Christiansen, NASA Johnson Space Center, Texas, United States

Dr. Albrecht de Jonge, SRON, The Netherlands

Dipl.-Ing. Peter Eichler, Technical University of Braunschweig, Germany

Dr. David Finkleman, United States Air Force Space Command, Colorado, United States

Dr. Vladimir Fortov, Research Center IVTAN, Russia

Dr. Edna Jenkins, United States Naval Space Command, Virginia, United States

Dr. Gennady Kuzin, NPO Energia, Russia

Mr. Joseph Loftus, Jr., NASA Johnson Space Center, Texas, United States

Ms. Lee Ann Hongping Lu, China Academy of Launch Vehicle Technology, People's Republic of China

Dr. Carl Maag, T&M Engineering, California, United States

Dr. Jean-Claude Mandeville, CERT-ONERA/DERTS, France

Dr. Darren McKnight, Kaman Sciences Corporation, Virginia, United States

Dr. Walter Naumann, ESA Headquarters, France

Mr. Robert Penny, Jr., Motorola, Arizona, United States

Lt Col John Rabins, Air Force Space Command, Colorado, United States

Dr. Robert Reynolds, Lockheed Engineering Services Center, Texas, United States

Mr. Lakkavalli Satyamurthy, Embassy of India, Washington, D.C., United States

Mr. Eugene Stansbery, NASA Johnson Space Center, Texas, United States

Mr. Hitoshi Takatsuka, National Space Development Agency, Japan

Mr. Jose Verissimo, Hughes Space and Communications Company, California, United States

Dr. R. Viswanathan, Hughes Space and Communications Company, California, United States

Prof. Menglun Yu, China Academy of Launch Vehicle Technology, People's Republic of China

Observers and Liaisons

Col Bill Gardner, Air Force Headquarters, United States

Mr. Russell Graves, Boeing, Texas, United States

Mr. George Levin, NASA Headquarters, Washington, D.C., United States

Mr. Duane McRuer, ASEB Chairman, California, United States

Dr. Walter Sarjeant, ASEB Committee on Space Station, New York, United States

List of Acronyms

AIAA	American Institute of Aeronautics and Astronautics
CCD	charge-coupled device
CIS	Commonwealth of Independent States
COBE	Cosmic Background Explorer
COPUOS	Committee on the Peaceful Uses of Outer Space
EOL	end of life
ESA	European Space Agency
EURECA	European Retrievable Carrier
GEO	geosynchronous Earth orbit
GOES	Geostationary Operational Environmental Satellite
GTO	geostationary transfer orbit
HAX	Haystack auxiliary radar
HEO	high Earth orbit
IAA	International Academy of Astronautics
IADC	Inter-Agency Space Debris Coordination Committee
ILA	International Law Association
LDEF	Long Duration Exposure Facility
LEO	low Earth orbit
LIDAR	light detection and ranging
MLI	multilayer insulation
MSX	Midcourse Space Experiment
NASA	National Aeronautics and Space Administration
NRC	National Research Council
ODERACS	Orbital Debris Radar Calibration Spheres
RCS	radar cross section

RORSAT   Radar Ocean Reconnaissance Satellite
SSC   Space Surveillance Center
SSN   Space Surveillance Network (USA)
SSS   Space Surveillance System (Russia)
UHF   ultrahigh frequency
UN   United Nations
VHF   very high frequency

Unit Conversions

Length

1 km (kilometer)	=	0.621 mile
1 m (meter)	=	3.28 feet
1 cm (centimeter)	=	0.394 inch
1 mm (millimeter)	=	0.0394 inch
1 μm (micron)	–	0.0000394 inch

Mass

1 kg (kilogram)	=	2.20 pounds
1 g (gram)	=	0.0353 ounce

Energy

1 J (joule)	=	0.239 calories
1 MJ (megajoule)	=	239,000 calories

Velocity

1 km/second	=	2,240 miles/h
1 m/second	=	2.24 miles/h

Force

1 N (newton)	=	0.225 pound (force)

Pressure

1 kbar	=	14,500 pounds/square inch

Glossary

APOGEE — point in an orbit that is furthest from the Earth.

BALLISTIC LIMIT — minimum thickness of a target (such as a debris shield) necessary to prevent an impacting particle from perforating it.

BREAKUP — destructive fragmentation of a space object. Breakups may be either accidental or intentional. Since the early 1960s, debris created by in-orbit breakups has represented the largest single constituent of the total space object population.

CATALOGING — process of detecting, identifying, and determining the discrete orbit of a space object. In cataloging, data from sensor networks are used to create a set of orbital elements that describe an object's discrete orbit. These orbital elements can be used to predict an object's future position, but must be updated periodically to account for orbital perturbations. Space object catalogs have been compiled and are maintained by different national governments and agencies.

DEBRIS — see "Orbital Debris."

DEBRIS FLUX — amount of debris passing through a given area in a given time. Area, as well as flux, can be defined in terms of either surface area or cross-sectional area. The debris flux experienced by a spacecraft is directly proportional to the probability of impact.

DECAY — natural loss of altitude of a space object culminating in reentry into the Earth's atmosphere. At low altitudes the rate of decay is determined largely by atmospheric density and the object's area-to-mass ratio, but for space objects in highly elliptical orbits, solar-lunar gravitational forces usually drive the rate of decay. Decay may be accelerated by lowering the perigee of an object's orbit.

DEORBIT — deliberate, forced reentry of a space object into the Earth's atmosphere by applying a retarding force, usually via a propulsion system.

FRAGMENTATION — process by which an orbiting space object disassociates and produces debris. Fragmentation includes such processes as breakup and physical deterioration due to exposure and aging. The planned, controlled, and intentional release of objects (see "Mission-related Object") is not considered fragmentation.

GEOSYNCHRONOUS EARTH ORBIT (GEO) — see entry under "Orbital Regions."

HYDROCODE — numerical computer capability to simulate hypervelocity impacts and the structural deformation, changes of state, fragmentation, etc., that result from such impacts.

HYPERVELOCITY — relative velocity of two objects that, in general, exceeds the speed of sound in solid materials (about 5 km/s) and results in an impact response that is not dominated by material strength effects.

INCLINATION — angle between the orbital plane of a space object and the plane of the Earth's equator.

LIGHT GAS GUN — two-stage gun device that uses a highly compressed light gas (such as hydrogen) to accelerate projectiles to typical speeds of 5-10 km/s under well-controlled conditions.

LOW EARTH ORBIT (LEO) — see entry under "Orbital Regions."

MISSION-RELATED OBJECT — object intentionally released from a spacecraft or rocket body during the course of a mission. These objects normally perform no useful service after release and are sometimes referred to as "operational debris." Examples of mission-related debris include spacecraft-launch vehicle separation and stabilization devices, sensor covers, and temporary protective shields. Debris from intentional breakups are not considered mission-related objects.

ORBITAL DEBRIS — space objects in Earth orbit that are not functional spacecraft. Spent rocket bodies, mission-related objects, fragments from breakups and deterioration, nonfunctional spacecraft, and aluminum particles from solid rocket exhaust are all considered debris.

ORBITAL LIFETIME REDUCTION — accelerating the natural decay of spacecraft and other space objects to reduce the time that they remain in orbit. Orbital lifetime reduction can be achieved through propulsive maneuvers, deployment of balloons or other drag-enhancing devices, and other methods.

ORBITAL REGIONS — Space objects travel in a wide variety of orbits at various altitudes. The following are some of the more frequently used orbits:

Low Earth Orbit (LEO) — orbit with a mean altitude of less than 2000 km.

Sun-Synchronous Orbit — retrograde LEO orbit in which the orbit plane precesses at the same rate the Earth revolves around the Sun. A spacecraft in SSO experiences the same ground lighting conditions each day; this can be useful for Earth observation missions.

High Earth Orbit (HEO) — any Earth orbit with a mean altitude greater than 2000 km.

Circular Semisynchronous Orbit — circular orbit (such as that used by the Global Positioning System) with a period of about 12 hours. The mean altitude of such an orbit is approximately 20,200 km.

Highly Elliptical Orbit — orbit with an eccentricity of greater than 0.5, including GTO and the Molniya orbits.

Geostationary Transfer Orbit (GTO) — elliptical orbit with an apogee around GEO and a perigee in LEO. This orbit is used to transfer spacecraft from LEO to GEO. The rocket bodies used to accomplish this transfer often remain in this orbit after the spacecraft separates and circularizes its orbit using an apogee kick motor.

Molniya Orbit — highly elliptical orbit with an inclination of 63-65 degrees, a period of about 12 hours, and an apogee above the Northern Hemisphere. Molniya orbits have historically been used to provide communications and early-warning services; they are suited to this task because spacecraft in Molniya orbits spend most of their time above the middle latitudes of the Northern Hemisphere.

Geostationary Earth Orbit — nearly circular orbit with a period of approximately 1,436 minutes and an inclination close to zero degrees. In such an orbit, the satellite maintains a relatively stable position directly above the equator, at a mean altitude of approximately 35,785 km. In practice, "geostationary" satellites exhibit small orbital eccentricities and slight inclinations, resulting in an apparent wobble about a fixed location.

Geosynchronous Earth Orbit (GEO) — roughly circular orbit with any inclination and a period of approximately 1,436 minutes. The ground tracks of inclined geosynchronous satellites follow a figure eight-shaped pattern, completing a full circuit once a day, with the center of the figure eight fixed directly above the equator at an altitude of 35,785 km.

PASSIVATION — discharging all stored energy sources on a space object in order to reduce the chance of breakup. Typical passivation measures include venting excess propellant and discharging batteries.

PERIGEE — point in an orbit that is closest to the Earth.

REORBIT — intentional changing of a space object's orbit at the end of its operational life. Typically, this involves putting the space object in an orbit where it is expected to be less of a hazard (including both collision and reentry hazards).

ROCKET BODY — any stage of a launch vehicle (including apogee kick motors) left in Earth orbit at the end of a spacecraft delivery sequence. Typical space missions leave only one rocket body in Earth orbit, but some launches leave as many as three separate rocket bodies in different orbits. Some rocket bodies may carry special devices for experimental purposes and be given names associated with the experiment. Rocket bodies are normally as large or larger than the spacecraft they carry and often retain residual propellants that may later be a source of energy for breakup.

SOLAR CYCLE ACTIVITY — periodic fluctuations in the energy output of the Sun. In general, these fluctuations exhibit an approximately sinusoidal variation with a period of 11 years. During periods of high solar activity, the Earth's atmosphere is heated, causing it to expand. This expansion increases the atmospheric density encountered by space objects, particularly those in orbits lower than 1,000 km, causing them to decay more rapidly. This may lead to a decrease in the overall population of objects in Earth orbit during solar maximum periods.

SPACECRAFT — orbiting object designed to perform a specific function or mission, (e.g., communications, navigation, or weather forecasting). A spacecraft that can no longer fulfill its intended mission is considered nonfunctional. (Spacecraft in reserve or standby modes awaiting possible reactivation are considered functional.)

SPACE OBJECT — any object in space. The term space object includes the natural meteoroid environment, as well as orbiting objects such as individual spacecraft, rocket bodies, fragmentation debris, and mission-related objects. It should be noted that the space law community has yet to come to consensus on the classification of debris as space objects.

SPACE SURVEILLANCE NETWORK (SSN) — collection of ground-based radar and electro-optical sensors used by the U.S. Space Command to track and correlate man-made space objects.

SPACE SURVEILLANCE SYSTEM (SSS) — Russian counterpart of the U.S. SSN. The SSS is located throughout the former Soviet Union and is comprised principally of radar, optical, and electro-optical sensors.

SPALLATION — phenomenon that occurs when a high-velocity impact causes a stress wave to interact with the free back surface of a thick target. If the resulting tensile stress caused by this interaction exceeds the tensile yield stress of the material, a thin sheet of material can separate from the target (or "spall") and be propelled from the surface at a velocity nearly equal to the original impact velocity of the particle producing the stress wave.

Index